SpringerBriefs in Computer Science

For further volumes:
http://www.springer.com/series/10028

Shaun Foster · David Halbstein

Integrating 3D Modeling, Photogrammetry and Design

Springer

Shaun Foster
Rochester Institute of Technology
Rochester, NY
USA

David Halbstein
Rochester Institute of Technology
Rochester, NY
USA

ISSN 2191-5768 ISSN 2191-5776 (electronic)
ISBN 978-1-4471-6328-2 ISBN 978-1-4471-6329-9 (eBook)
DOI 10.1007/978-1-4471-6329-9
Springer London Heidelberg New York Dordrecht

Library of Congress Control Number: 2013958143

Printed on acid-free paper

Springer is part of Springer Science+Business Media (www.springer.com)

Preface

This book was inspired by observations, research over the last several years, and subsequent talk on the subject at SIGGRAPH 2012. The field of 3D Digital Design has added several photogrammetry and retopology-based technologies resulting in many new and accelerated workflows, which have also affected the 3D design education process.

It has been a journey not only in the writing, but in the research and approach. What we describe in these pages is an ongoing case study that started many years ago, and will continue for as long as we continue to teach and to learn. As "early adopters" in the field of technology influenced design-based education we naturally push for advanced technology, but as part of a larger institution we must accept what the constraints of budget and adherence to tradition dictate. We have found that these opposing forces create a natural balance; that in some cases constraints lead to greater creativity than freedom ever can—but in other cases the opposite is true.

The book looks at the convergent nature of technology and its relationship to the field of photogrammetry, and 3D design. This is a facet of a broader discussion of the nature of technology itself and the relationship of technology to art, as well as an examination of the educational process itself. We have addressed the question of how we are adjusting and will continue to adapt to further disruptive technologies.

Over the years, we have been faced with this question and more; in the development of a post-secondary curriculum we are always questioning—are we teaching art and design, or are we training technology? Are we teaching our students to think, to ideate, to create, or are we teaching them how to use tools? Clearly, the two concepts are inexorably intertwined, but in any educational institution or college curriculum it is important to make the distinction, and to understand the areas that are clearly defined as well as those that are not.

Each year we are presented with new technology. Software and hardware tools are developing at an extremely rapid rate and with each new change, each new outgrowth, we must make important decisions about whether or not these new tools are important enough to incorporate into our existing curriculum. Do we keep pace for the sake of keeping pace, or will our adaptation to change foster creativity and new thought? Given the existing constraints of time and budget, what are we willing to sacrifice in order to embrace these changes and fold them into our curriculum? Which tools, methods, approaches, or classes can give way to this change? Clearly these questions seek answers, this book offers insights for ways to integrate some of these new technologies into the field of design, and from a broader standpoint it also looks ahead, raising further questions and looking to the near future as to what additional technologies might cause further disruptions to 3D design as well as wonderful creative opportunities.

Contents

Chapter 1
Introduction

Abstract The introductory chapter opens the book by giving a brief overview of the main areas of discussion: An introduction to photogrammetry, computer vision, 3D modeling, surface topology, and retopology and finishes with a brief discussion of the design process.

Keywords Introduction · Photogrammetry · Computervision · design process · Retapology · 3d design · Education · Entertainment field

1.1 Introduction

The focus of this book is to identify and discuss the rapid changes to 3D computer modeling workflows, as represented by the integration of photogrammetry, high-polygon density-interactive modeling, and retopology software capabilities. We recognize that these approaches have been part of the entertainment and broader "visualization/remote sensing industries" for many years; however, our focus is on the accessibility in the field of design and an examination of the impact these technologies will have on the field of design as well as the pedagogical nature of 3D design. As these advanced capabilities become less expensive, more refined, and therefore more available to secondary and post-secondary educational institutions, it is incumbent upon those institutions to address their integration into a changing curriculum and recognize the changes these approaches represent to a pedagogic paradigm. In addition, we will also give a historical context for the tools for the fields these technologies originated in using this as a basis to briefly discuss technological convergence as a trend between multiple fields, and to the relationship it has to elements and principles of design in addition to considering what further aesthetic and technical convergences might be on the horizon.

Therefore, within the pages of this book, we will discuss modeling methodologies that are currently being taught, explore their advantages and disadvantages, and offer conjecture as to how this field is changing from the perspective of education and

S. Foster and D. Halbstein, *Integrating 3D Modeling, Photogrammetry and Design*,
SpringerBriefs in Computer Science, DOI: 10.1007/978-1-4471-6329-9_1,

curriculum development. The tools, workflows, and art of generating 3D computer models are expanding and rapidly changing. Over the past 20 years, most 3D objects have been created by highly trained, technical 3D artists through a labor-intensive process. This usually starts with simple object, adding complexity until the desired form is reached. A small minority of 3D objects were created using scanning 3D technology which has not gained broad use due to high cost and also difficulty of using the final product generated by laser scan. The 3D pieces generated by these scans were "heavy" on geometry, starting in the hundreds of thousands to millions of polygons which make them inefficient to modify in some way, for example a high-polygon object will take much more processing power to bend, twist, attach to other objects, and also there is higher computer costs for rendering, the transformation process of turning 3D objects into 2D still images.

Several maturing technological components are combining which open a very different workflow, reducing the costs requirements of human labor and output of heavy geometry. The first component is photogrammetry technology, which is the process of using photography to determine geometric measurements of objects. The second are several technologies from the field of computer vision. This field is applying computer algorithms that automate the understanding and output from photogrammetry. Finally, the third component is "retopologizing." New tools in restructuring geometry into clean and efficient forms have the potential to instantly create 3D objects and optimize their geometry.

As in any emerging technology-dependent medium, the tools used to create visual images have a great impact on the final output. From an aesthetic point of view this can be both unintentional and undesirable. Rather than allow the limitations of technology to drive our artistic goals, as artists we seek the reverse, we aim to enable our artistic vision to define the tools we use. Within this context, we will survey the main components of design and investigate how going from an additive process of constructing 3D objects (which is currently a major component of university curricula in this discipline) to a subtractive process (as represented by the new technology discussed in this book) potentially changes what people design, by freeing them from the constraints of technical limitations. We also discuss where we see new opening areas for creative process as well as potential pitfalls for digital artists.

This book concludes with a list of online resources, many of which are free, and outlines ways of delivering structured content for those interested in learning and participating in building 3D objects, manipulating them, and using them in 3D visualizations or even outputting them to 3D printers.

1.2 An Introduction to Photogrammetry

Photogrammetry is the science of extracting reliable measurements from two-dimensional (2D) images, usually photographic. This field of study has a history which connects it to many other fields and disciplines. Some of the disciplines include: optics, projective geometry, remote sensing, and more recently computer vision. From a technical standpoint, we are mainly focused on the output of three-dimensional (3D) meshes generated from a workflow utilizing a more sophisticated technique of photogrammetry known as stereophotogrammetry. Stereophotogrammetry involves estimating 3D coordinates of an object by comparing multiple photographic images taken from different positions. From these multiple images, a ray (or line) can be calculated to a 3D point(s) on an object.

More about the history and basics of photogrammetry field will be discussed in Chap. 2.

1.3 Computer Vision

This field is concerned with the process of acquiring, processing, and analyzing images with an overall goal being to create "useful" information. Computer understanding of images in a way that can produce multiple, usable, different types of information can then be applied in many ways. Some applications are similar to vision-related tasks which humans currently perform, such as inspecting products on an assembly line; other tasks transcend human tasks by "seeing" beyond the light spectrum visible to humans, X-rays, and infrared. Similar to photogrammetry, this field is at the heart of multiple other emerging technologies; artificial intelligence, robotics, machine learning, psychology, and many more traditional fields such as mathematics and physics. The discussion in this book of the computer vision field will be constrained to its connection to photogrammetry. However, some discussion will be given to speculation of ways where emerging technologies being developed by computer vision might affect 3D design in the future.

1.4 Surface Topology and Retopologizing Workflows

3D computer graphics for film and video games use surface modeling techniques to generate 3D "meshes." The 3D model is comprised of a topology of connected points that connect to form the geometry which describes the surface of the object (Fig. 1.1). Surface modeling is mainly concerned with describing the 3D object for use in image creation. Depending on the complexity of the surface, a high amount of detailed geometry may be necessary. The usual process of surface 3D modeling entails building from a simple form and adding complexity and details. In this book, we will address several methods for building 3D objects as well as define surface versus solid modeling. In order to find a balance between an over complex geometric surface and one that does not provide enough detail, "retopologizing" is a process of optimizing 3D mesh geometry by rebuilding it (Fig. 1.2). The number of polygons is reduced over 200 %, while the overall form remains recognizable, reducing resources needed by the computer.

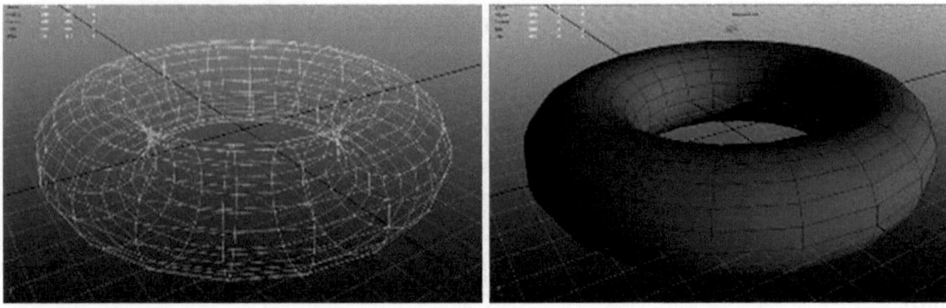

Fig. 1.1 3D Mesh. Autodesk screen shots reprinted with the permission of Autodesk, Inc

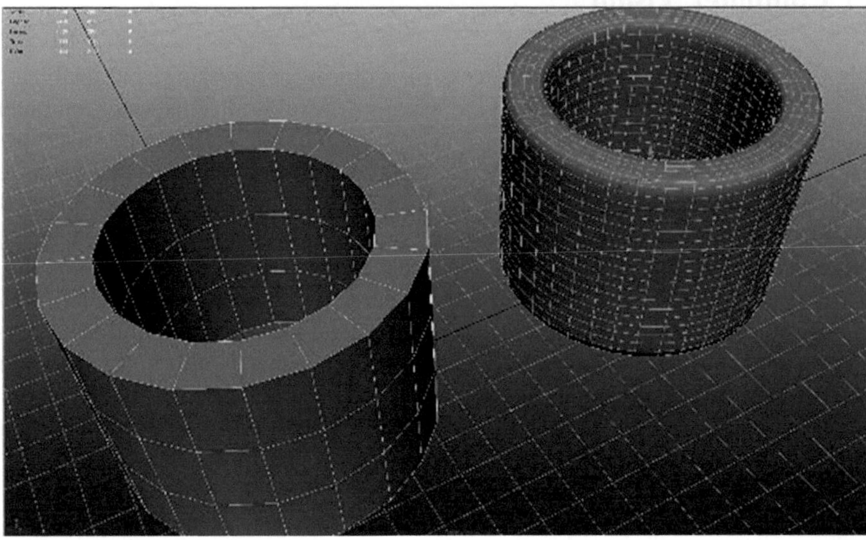

Fig. 1.2 Retopologization. Autodesk screen shot (Maya) reprinted with the permission of Autodesk, Inc

1.5 Design Process

While many designers work to avoid the "medium" becoming the message, the technical and creative process for planning and creating can be highly influenced by the tools used. In this book, we will discuss many of the creative avenues and design possibilities that are opened by combining new tools in ways that are just beginning to be explored. Many of the elements and principles of design connect in various ways to the process of 3D form as well as perception of objects.

Chapter 2
Technology Leaders and Adopters

Abstract This chapter gives an overview of the nature of technology and investigates many of the other technologies which are combined for photogrammetry to emerge in its current state. From general technology theory to a brief history of photogrammetry, it investigates some of the technologies which incorporate photogrammetry technologies and are also connected to the fields of design and entertainment: Microsoft Kinect, photosynth, Weta Digital: Photospace, Autodesk 123D Catch. This chapter concludes by investigating some of the current challenges and limitations.

Keywords Major technology · Convergence · Future studies · Technology cycles · Art and design · Photogrammetry history · Nature of technology · Photogrammetry · Microsoft Kinect · Photosynth · Kinect fusion · What a digital · Photos space · Autodesk 123D catch · Challenges · Limitations

2.1 The Nature of Technological Spread

Technology has evolved in alongside human societies for the last several thousand years. While multiple definitions exist, a combination of its qualities shows it to be a continuously evolving process of harnessing and amplifying sources of power which were originally part of the natural world in the form of tools or processes which serve to increase human abilities to fulfill our needs and desires [1]. Two components of the nature of technology are being addressed here: the nature of technology and accelerating change.

The nature of technology reflects similarities to the evolutionary nature of biological organisms. There are evolutionary cycles in technology similar to biological development; many new technologies are tested, and the most useful combinations move ahead. The combinations often start off somewhat weak (birth) but then (grow) strengthen and optimize (mature) over time. As the technologies reach maturity, there is often a need to combine with additional technologies in order to spawn a new cycle in order to avoid disuse (decline).

S. Foster and D. Halbstein, *Integrating 3D Modeling, Photogrammetry and Design*, SpringerBriefs in Computer Science, DOI: 10.1007/978-1-4471-6329-9_2,

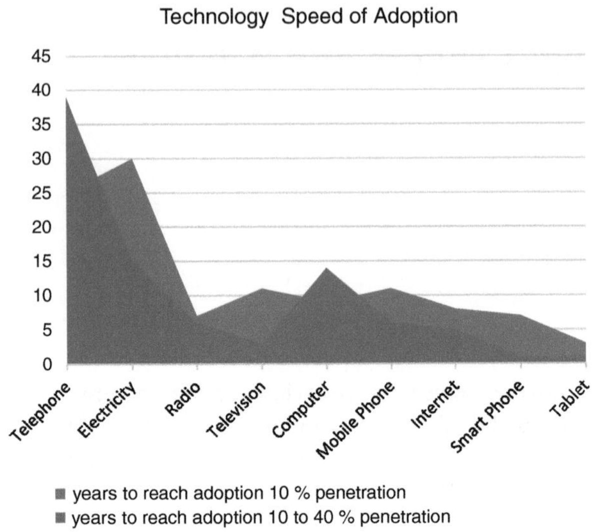

Fig. 2.1 Mass use of inventions [Technology Review, ITU, New York Times, WSJ, US Census Bureau]

The field of future studies has described accelerating change as the perception of increased numbers of technological changes over the last 2,000 years of history. There has been a growing amount of statistical research showing a logarithmic trend of major technological change, combined with a reduction in the time a society takes to adopt the technology. Figure 2.1 presents several recent technologies and their speed of adoption [2]. A simple but controversial view connects Moore's Law, which describes the logarithmic growth of transistors in microprocessors, to accelerating technological progress [3]. While there is some controversy as to exact formula, others work to present more accurate ones [4]. Some of the factors linked for new technology adoption appear to be costly and technically difficult to use.

The result of the accelerating change has been perceived as "disruptions" to many fields, which had been stable. The disruptive technologies have at times completely destroyed (or radically changed) whole industries.

Over the last 20 years, 3D computer graphics has been a rapidly moving, growing, and evolving field. Many of the technical challenges within the field have been solved. The creation of 3D objects has been in high demand for use in multiple fields, from visual effects, animated films, architecture, video games, etc. Creating those objects has been done by the briefly described process of 3D modeling. Over time, those tools have improved and an industry has developed around building necessary 3D objects. While there have been incremental improvements in the tools and processes used for creating 3D objects, the field of 3D modeling had a similarity to skilled craftsmen's guild, due to the mix of high technical skill mixed with artistic knowledge. Given the pace of technology, it seemed inevitable for this field to have a disruptive change. This change is in the

Technology	Approximate Dates
Photography & plane tables	1850-1900
Stereoplotters & Airplanes	1901-1950
Computers & Mathematical Models	1951-1971
Digital & Computer Vision	1972-Present

Fig. 2.2 Stages of photogrammetry development

integration of photogrammetry (using real images and video technology) and procedural object creation (objects made by the use of pre-programmed algorithms). In this book, we will be focusing on the emerging workflow that involves photogrammetry and retopologization.

2.2 A Further Introduction and History of Photogrammetry

Photogrammetry relies on gathering data from images that can be analyzed to extract and yield further information. It has an extensive history, with origins connected to mathematics, optics, and geometry, which can be traced as far back as 1492, with DiVinci describing the principles of optical perspectivity [5]. Over time, its first broad application was to aid in the creation of highly detailed cartography. As the technology progressed, the data that photogrammetry and remote sensing equipment are capable of acquiring have expanded to physical, temporal, and semantic types of information ([8], p. 4). The field, however, still has strong connection to its origins. Photogrammetry falls under the field of "remote sensing." This is because while it is doing advanced calculations and analytic measurement of objects, there is no physical contact during analysis.

In the scientific fields, there are sometimes additional categorizations which place the remote sensing aspect of photogrammetry as information only from imagery which is received from satellites (images from above the earth). For the sake of this book, we will mostly discuss photogrammetry as it is being applied to the creation of 3D objects and will not be addressing the scientific and mathematical components of remote sensing.

The history of photogrammetry can be separated into four close differing stages based on improvements in technology. The stages, titles, and dates vary slightly by different scholars ([6, 7], p. 282) and ([8], p. 7); however, the general stages of development can be separated into the following approximated time periods (Fig. 2.2).

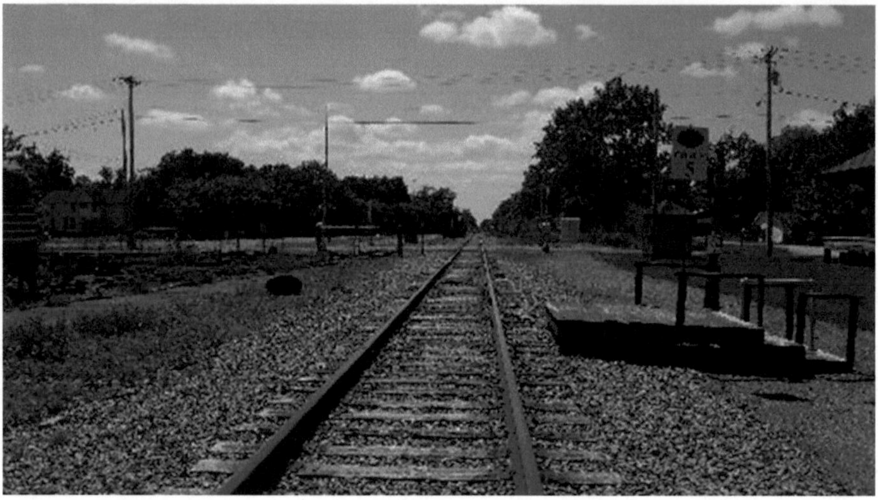

Fig. 2.3 Perspective can be generated using Euclidean geometry

2.2.1 Photography and Plane Tables

The current state of photogrammetry required many technological advancements over the last several hundred years. Mathematics and optics were critical initial components, since photogrammetry relies on the scientific analysis of photographs. As previously mentioned, the history of photogrammetry in the West can be traced back to as far as the fifteenth century [10] (Italian Renaissance). Rules of linear perspective can be derived from the basic Euclidean geometry. To this day, artists use underlying principles of Euclidean geometry within their drawings to create realistic 3D representations of the world. The translation from real-world 3D image, can be done “by eye” seen from real life, or using more exact processes such as camera obscura images projected onto a grid, or more recently images taken with a digital camera which can then be traced. The goal of photogrammetry is to do the opposite: to use mathematics to extract information from images.

Photography and plane table analysis was performed on images taken from high rooftops, from hills, or from balloons or even using kites and pigeons. Once these images were taken, linear perspective combined with simple math was able to manually compare the known heights in order to derive additional highs of objects within the image (Fig. 2.3).

2.2.2 Stereoplotters and Airplanes

Stereoplotters are stereoscopic images combined with technologies for automating the measurements of objects from two overlapping and optically corrected images. There were multiple inventions (floating mark, stereocomparator, stereoautographs, serial-photo aerial camera, etc.), which began to automate height analysis of images in the early twentieth century. By the 1930s, advances and refinements of these technologies as well as improvements in aviation resulted in the use of aerial stereophotography becoming the main method of map making.

2.2.3 Computers and Mathematical Models

The further refinement of analog devices for stereoplotting and photography was combined with the rapid calculations which are able to be performed by computers allowed for even greater accuracy. Specifically, some of the more complex "algorithms for orientation and triangulation..." were developed, dramatically accelerating results and accuracy.

2.2.4 Digital and Computer Vision

The first readable images of Earth sent from orbit were from Landsat in 1972. The transition from partial to near-complete automation has taken another 40 years. Digital photography, further improvements in algorithms, as well as the ability to record in multiple formats allow for greater interpretations. "Pan-chromic imagery, near-infrared, and color" could all be taken simultaneously using digital, removing the need for chemical development of images and scanning. Also, many of the physical plotting devices such as the stereoplotters have been replaced by lIght and range detecting (LIDAR) remote sensing technology that uses lasers instead of analglyphic stereoimagery to map depth and distance information.

2.2.5 Looking Ahead

When it started, the field was used for extremely large mapping applications, but in the near future, a microscopy technique using stereophotogrammetry is being planned for the analysis and creation of ultrasmall components [9]. The technology is also being used to look and map areas farther away. Currently LIDAR, a remote sensing technology, which combines laser and radar technology, is mapping the moon. Many of the photogrammetry techniques are beginning to move into other research fields as well as become accessible to mainstream population for use in creative, educational, or hobbyist application.

2.2.6 The Cycle Between Art and Science

A current list [6] of applications of photogrammetry is

- Mapping and acquisition of geoinformation
- Documentation
- Monument preservation and architecture
- Aerial, terrestrial, and underwater archaeology
- Monitoring earth surface and building deformations
- Civil engineering studies
- Automobile, aeronautical, and nautical industries
- Dental, orthopedic medicine and biomechanics
- Forensic applications.

This book specifically deals with some of the ways a new category is being added to this list:

The fields of art and entertainment, specifically how new technologies integrate into college 3D art and design education. Interestingly, in the past, artists adopted mathematics and linear perspective to derive 2D images that look 3D, and now, the current field of 3D computer graphics and digital design is once again connecting with the fields of science (photogrammetry) and full cycle, by adopting photogrammetry technology to derive 3D objects into artwork. The field of 3D computer graphics has begun using photogrammetry information to rapidly generate 3D models, in order to make visual images based on real life. While there have been various market leaders in industry who have sparsely applied these technologies, the factors of low cost and mass accessibility are laying the groundwork for broad adoption.

As the cycle continues of art, science, and technology, we will now introduce several of the applications created by the entertainment industry for use in research and application.

2.3 Photogrammetry Resources Connected to the Entertainment Industry

Photogrammetry for deriving 3D assets (and analysis) has existed in many forms and in various products within other fields. Within the design and technology fields, the technology has arrived relatively recently. In the next few sections, we will focus on a few technologies developed by Microsoft, Autodesk, and Weta Digital , due to their connections to the media and design fields. However, we have also compiled a list of additional photogrammetry resources in the appendix of this book.

2.4 Microsoft Kinect

Due to its industry-leading research laboratories in computer vision combined with its Xbox and Kinect video game platform, Microsoft has been a market leader with combining photogrammetry and computer vision technology. The Kinect, a breakthrough device, was launched as a peripheral to the Xbox 360 game system in November 2010. However, Microsoft (and others) launched additional software development kit (SDKs) which allowed the device to be used in ways beyond its intended use as a next-generation game controller.

The Kinect device combines a standard camera, infrared projector, and microphone. By analyzing visual information generated by the camera and depth information generated by the infrared sensor, it creates a large set of ways to interact with computers without a mouse or keyboard. Gesture-based interactive technology in the Kinect was acquired by Microsoft from a range camera developed by PrimeSense, an Israeli 3D sensing company. Beyond 3D sensing, the new 3D and gestural interface has had a large impact on the field of user interface design as well (Fig. 2.4).

In May 2013, **Xbox One** was announced by Microsoft, and the upgraded "Kinect One" sensor is due out with the release of the gaming system late 2013. With this product will come further improvements to the Kinect Fusion technology.

The Kinect One hardware will feature a big step forward in technology. It incorporates greater gesture recognition, up to six people tracking (via facial recognition) and a full HD 1080P resolution (as compared to the original sensor which offers only standard definition). It also promises to deliver a nearly latency-free workflow.

Probably, the most significant feature listed above in respect to this book is the nearly 4X increase in resolution and speed of capture. 3D computer models, which will be generated by the Kinect One, will have greater detail for more accuracy. This additional fidelity will make capturing 3D models much more accurate.

2.5 Microsoft (CV) Kinect Fusion

First presented at SIGGRAPH 2011, this technology is a series of software libraries that connect to the Kinect hardware. Kinect Fusion was developed at the Microsoft Computer Vision Laboratories. One component of the Kinect Fusion technology allows for high-quality, 3D renderings of environments and people in real time. Most current photogrammetry to 3D mesh workflows involves taking

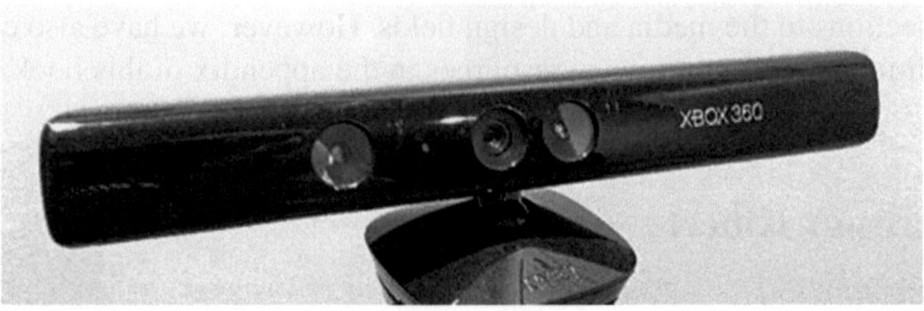

Fig. 2.4 Microsoft Kinect sensor

multiple photographs and then having comparatively analyzed in order to extrapolate 3D point cloud volumes, from which 3D meshes are derived. Kinect Fusion develops 3D meshes in real time. We anticipate faster and more accurate detailed meshes that will rapidly emerge as the technology further evolves. Other products have also incorporated 3D the Kinect as a platform for scanning technologies (for example, http://reconstructme.net/).

In March 2013, Microsoft released a significant update to the Kinect SDK. This update **includes** the Kinect Fusion technology in the SDK feature set (Microsoft Kinect SDK) [11].

- Real-time, GPU-assisted 3D object and scene reconstruction by using the Kinect for Windows sensor
- Ability to infer relative sensor position and orientation from a 3D scene for augmented reality application
- Advanced algorithms that are powerful enough for large sensor movements and scene changes during scanning
- Direct X11 compatible graphics cards supported
- AMD Radeon 7950 and NVidia GTX560 have been validated to run at interactive rates
- Kinect Fusion Studio and samples demonstrate 3D scanning capabilities
- Non-real time CPU mode for non-interactive rate scenarios.

2.6 Photosynth

This product allows for two different types of visualization based on photogrammetry technologies. It was developed in collaboration between University of Washington and Microsoft research laboratories. There are currently two functions within Photosynth, Synths, and panoramas. The "synth" uses multiple images that are analyzed in order to generate a three-dimensional image of the space. The second is a panorama which allows the user to take multiple pictures in a three-dimensional space. These pictures are then processed. The processing analyzes the images and combines them in a process called stitching. Figure 2.5 demonstrates

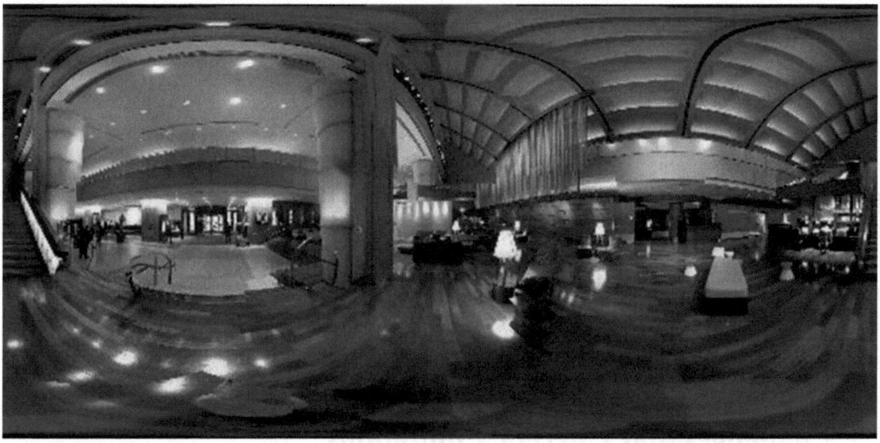

Fig. 2.5 Stitched image

the output of a stitched image, warped in a way so that it can be mapped to a spherical interactive QuickTime VR environment. This image was created using Photosynth, a product developed by Microsoft computer vision research laboratories.

There are a large number of Photosynth experiences is available on the Microsoft Web site: http://photosynth.net/explore.aspx. In addition to giving a description and showing the date created, many of them are geotagged (GPS info embedded in the metadata) for easy connection to map information.

2.7 Weta Digital: Photospace

In their presentation and published paper [12], they discuss the integration of a photogrammetry-based workflow for physical props for digitization and use in visual effects sequences. In their paper, they discuss previous workflows which use photographic reference, 3D modeling and 3D scanners. They describe the photogrammetry (vision-based) approach as combining the best features of multiple approaches. They described the workflow process as having three parts: (1) the capture session, (2) photogrammetry processing session, and (3) reference generation session. Images are captured, they are processed, and then, the three-dimensional models that are generated are handed to 3D artists for retopologizing or remodeling in order to make them efficient for use in the rendering pipeline. They also discussed many of the challenges and limitations connected to other photogrammetry-based workflows (see below).

2.8 Autodesk 123D Catch

123D Catch is a free photogrammetry software tool created by Autodesk Corporation. It has been produced for online use, in mobile device "app" form as well as standard desktop. The simplicity and quality of output generated by this software have made it very popular. Creating 3D meshes from photography is a simple process. The user takes 20–40 images of an object (with a maximum of 70) and feeds them to the program, which develops a 3D mesh from an analysis of the images. The software finds and matches common features in order to construct a 3D mesh from the identified feature sets.

2.9 Challenges and Current Limitations

While there have been many improvements in both speed and accuracy over the last 20 years, there are still hurdles to overcome. Many of the algorithms that are used for translating computer vision resources into 3D mesh objects can be easily broken, which generates a failed, problematic, or incomplete 3D mesh. These limitations listed below are fairly standard among the previously listed and most other current photogrammetry technologies. However, most softwares list these as heuristics to overcome many of the issues with current programs, because knowing the use guidelines can help overcome failed or problematic meshes.

2.9.1 Occlusions and Number of Photographs Necessary

Occlusions can cause problems. An occlusion is when and unwanted object comes between the camera and the target photogrammetry object. In small scenes, this can be things like imaging a human who wears glasses (also see below regarding transparent objects). An example of a similar problem for a large exterior scene would be a tree in front of a house. Most semicomplex objects have moderate self-occlusion, especially if they have multiple folding or complex parts. However, with enough images from the correct angle, many current photogrammetry algorithms are robust enough to solve for the objects. For heavy occlusion, images every 5–10°, overlapping as much as 50 %, might be necessary. For object with little or no occlusion, images ever 20° or more can be used for excellent results.

2.9.2 Photographs Need Features

The algorithm searches for parallax shifts between known features within multiple images. Taking images of blank walls or large empty/non-focused areas will cause the algorithm to fail. Patterns, strong lines, and differentiated features are what the algorithm looks for when tracking. Sometimes adding features with tape, stickers, or draped cloth can get better results. Similar to objects with no features (like the previously discussed blank wall), repetition of features can cause a similar confusion for the algorithm when trying to match feature sets.

2.9.3 No Transparent, Reflective, or Glossy Subjects

This creates certain difficulties for many objects. Even semitransparent objects can pose problems. When applicable, spray-painting them with matte finish can remove any transparent or shiny qualities. However, for large objects, for example buildings with reflective and semitransparent windows, they can pose difficulties.

2.9.4 Subjects Cannot Move During the Image Capture Process

In order for the algorithm to match the feature sets, object cannot significantly move.

2.9.5 Consistent Lighting

This not only means not changing light sources, but the algorithm works best **without** strong directional sources of light. Meaning outdoors, cloudy days (ambient light) would create the best lighting conditions. Indoors might require diffusers to create as ambient and consistent lighting conditions as possible. Additionally, the use of a flash will cause problems due to the fact it will create a unique (directional) lighting situation for each image.

For further reference listing the limitations of the 123D Catch, please follow this link to the following video: http://www.youtube.com/watch?v=7TfXXJxDsXw#at=64.

References

1. Arthur, W.B.: The Nature of Technology: What it is and How it Evolves. Free Press, New York (2009)
2. Degusta, M.: Are Smart Phones Spreading Faster than any Technology in Human History? Technology Review: Published by MIT http://www.technologyreview.com/business/40321/ (2012)

3. http://www.kurzweilai.net
4. Hamasher, K.: Accelerating Changes in our Epoch and the Role of Time-Horizons, pp. 205–215. Springer Netherlands, Heidelberg (2006)
5. Ghosh, S.K.: Fundamentals of Computational Photogrammetry. Concept Publishing Company, New Delhi (2005)
6. Redweik, P.: Photogrammetry. In: Xu, G. (ed.) Sciences of Geodesy—II: Innovations and Future Developments (Chap. 4), pp. 133–183. Springer, Heidelberg (2013). ISBN 3-642-27999-6, 978-3-642-27999-7
7. Konecny, G.: Geoinformation: Remote Sensing Photogrammetry and Geographic Information Systems, 2nd edn, p. 280. CRC Press, Boca Raton (2002)
8. Schenk, T.: Introduction to photogrammetry. Department of Civil and Environment Engineering and Geodetic Science, The Ohio State University, Columbus (2005)
9. Boutin, C.: Microscopy Technique could Help Computer Industry Develop 3D Components. NIST Tech Beat. http://www.nist.gov/pml/div683/tmsom-062513.cfm June 2013
10. Gruner, H.: Photogrammetry: 1776–1976. Photogram. Eng. Remote Sens. **43**(5), 569–574 (1977)
11. Microsoft Kinect SDK http://www.microsoft.com/en-us/kinectforwindows/
12. Bhat, P., Burke, S.: Weta digital PhotoSpace: a vision based approach for digitizing props, ACM SIGGRAPH 2011 Talks Article No. 1, New York (2011)

Chapter 3
Conceptualizing Basic 3D Modeling Workflows

Abstract This chapter traces the history of the development of the methods and practices used in 3D modeling. It also draws parallels between this development and its relationship to sculpture and design. In its various sections, a detailed discussion of the various approaches is provided; illustrating the changes in method as hardware and software has become more sophisticated over the years. It demonstrates how an artist uses existing and new technology to achieve aesthetic goals and how technology has been influenced by this achievement.

Keywords Additive · Subtractive · Modeling · Topology · Workflows · NURBS · Patch modeling · Surface parameterization · Polygons · Box modeling · Edge flow · Normal maps · UV layout · Langer · Retopology · Maya · 3D Coat · Mudbox · Zbrush · Digital sculpture

3.1 Compartmentalizing the Learning Curve

In teaching 3D modeling in general, and more specifically organic 3D modeling in a post-secondary environment, we find that there are two distinct areas of ability among our students, the aesthetic and the technical. Some of our students have developed skills in one area over another—as a result of personal interest, program availability in their high school, or other circumstantial or motivational reasons.

We can, of course, rate these skills using a very general scale of low–high in either area, which creates the following metric:

S. Foster and D. Halbstein, *Integrating 3D Modeling, Photogrammetry and Design*, SpringerBriefs in Computer Science, DOI: 10.1007/978-1-4471-6329-9_3,

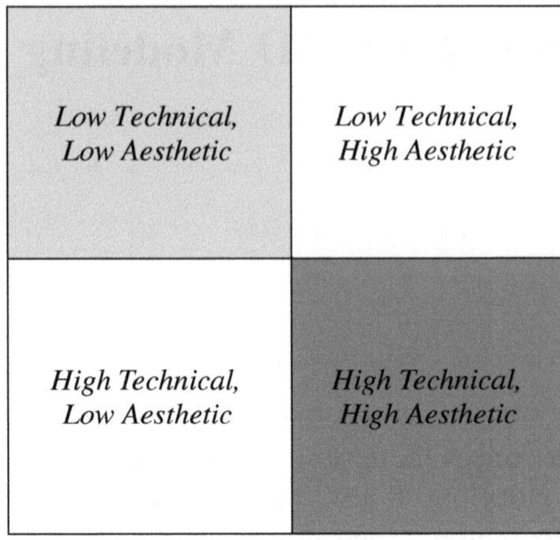

Of course, we want to build on whatever skills the student has in either area, while creating a foundation in the complementary area. Our goal is to develop our students from both sides, concurrently providing a highly developed aesthetic, strong sense of design, understanding of advanced principles of 2D and 3D design, and a well-cultivated understanding of anatomy on one side, and a mastery of the technology from both a procedural and conceptual approach, as represented by the following Venn diagram:

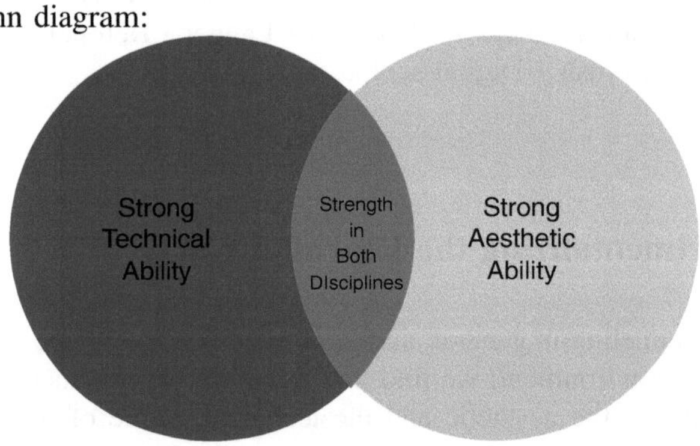

Separation

In any kind of process that uses technology or technical methodologies in the creation of art, the artist has to master the craft as a necessary means to an end. Wood grain, metallurgy, chemistry—all are part of the technical training for designers, sculptors, ceramicists. Without an understanding of the technology and craftsmanship involved in the structural integrity of a piece, the best designs will fall apart, or will be unusable. 3D modeling is no different.

In many disciplines, students are asked to design on paper first, without regard to the details of material or structural integrity. The approach taken by many instructors is "You design it first—then we will figure out how to put it together." Because of the laws of physics, material science, structural engineering—this is frequently the only viable approach.

As in any other discipline, it has been difficult if not outright impossible to separate the technological processes from the aesthetic development while using 3D modeling software. Understanding of edge placement, "flow" of polygons or patches, limitations on acceptable geometric forms ("quads" vs. triangles, non-manifold geometry, edge quality, etc.) is vital to the most basic kind of construction and presents a serious impediment to the creation of 3D art.

However, these barriers have come down almost entirely, thanks to the parallel development of three distinct types of software.

The first of these is digital sculpture software. The most commonly used digital sculpture applications today are Pixologic "ZBrush," Autodesk "Mudbox," and "3D-Coat," but there are at least a dozen more that are available as standalone packages, plug-ins, or as integrated modules in well-known professional and consumer while each package offers unique toolsets and focuses on different workflows and capabilities, all have the capability of "voxel sculpting." Voxel sculpting, simply put, is the process of using specific toolsets to push, pull, smooth, grab, pinch, or otherwise manipulate a digital object as though it were made of a real-life substance such as clay.

Voxel sculpting represents a departure from traditional modeling methods in that it can rely entirely on high-polygon meshes for detail. The old rules of clean topology will always apply in the end, but the methodologies outlined in this chapter will focus on the "brute force" capabilities of voxel sculpting without regard to those rules. This brute force approach is used in order to maintain the separation between purely interactive sculpture in a very traditional sense and the attention to the topology details which will be addressed in a second pass.

The second type of software is called "photogrammetry" and would be used instead of the voxel-sculpting described above. Photogrammetry also produces a high-density mesh that achieves detail by brute force, increasing density where additional detail is needed with no regard for model efficiency. Rather than use interactive sculpting tools, however, the process of photogrammetry uses photographic data and various triangulation algorithms to determine the geometric properties of the subject. For our purposes, photogrammetry software enables us to take a series of photographs of a single object from various points of view and derive a dense, but highly accurate, 3D mesh from those photographs.

The third type of software, and the one that makes the first two viable modeling options in the first place, deals with retopology. Retopology is the process of taking a high-density mesh and tracing it with new polygons in order to reconstruct the original using less dense, more efficient topology. Using this process on a pre-sculpted model allows the user to focus entirely on issues of edge placement, edge flow, and efficient geometry, with no additional regard to the sculptural aesthetics of the piece.

The workflow, therefore, would begin with the acquisition of an accurate, though overly dense, 3D mesh via interactive sculpture or photogrammetric capture, followed by a retopology process to create an efficient, lower-density mesh with accurate and systematically contrived edge placement.

Once the retopologized mesh has been created, it can be reacquired by the voxel-sculpting software for additional sculptural details, proper UV layout, and the subsequent generation of normal and displacement maps, which will be discussed later in this chapter.

3.2 3D Modeling: A General Conceptual Approach

Throughout the short history of 3D modeling, the level of complexity of 3D models has been directly related to the graphics capability of workstation hardware. This relationship has historically been a limitation, and it is this limitation that directed the development of commonly accepted modeling methods that we refer to as "Additive" workflows.

Whether using NURBS or polygonal modeling workflows, the additive methods all start with a simple form which is built up stepwise line by line, edge by edge, until the final detailed model is achieved. This method has usually been deemed necessary and proper, needed in order to prevent a work-in-progress from becoming unmanageably dense. To this end, particular attention is paid to edge placement and edge flow, both of which act as character lines necessary to define large shapes.

For the examples of modeling methods illustrated below, we have chosen the human head as the subject. The human head requires a substantial amount of attention to a relatively high degree of detail in order to accurately express the underlying musculature, the general morphology and the character of the person being modeled. For models intended for animation, the edge flow is doubly important, as the placement of these edges will be the lines defining the folds, wrinkles, and muscle masses as facial expressions change over time.

Regardless of the technological approach used, we believe it is vital to begin our modeling instruction with a study of the shape and forms of the subject, including the underlying structures—muscles, tendons, cartilage, and bones. The layout of these structures, their character lines and flow-lines, will help us to determine our edge placement when we begin modeling in our 3D software (Fig. 3.1).

In Fig. 3.2, one student begins by comparing an image of facial anatomy from an anatomy text to the head she wants to model. Using this as a reference, she then identifies important "Landing Zones," or skeletal ridges, volumes, edges, and masses that contribute to the overall shape of the head. These skeletal structures are key in defining the origin and insertion locations of various muscle groups and help to begin the strategy of edge flow.

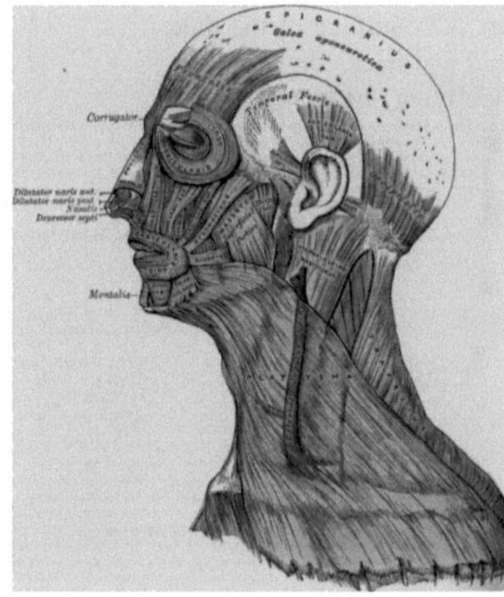

Fig. 3.1 Muscles of the face and neck

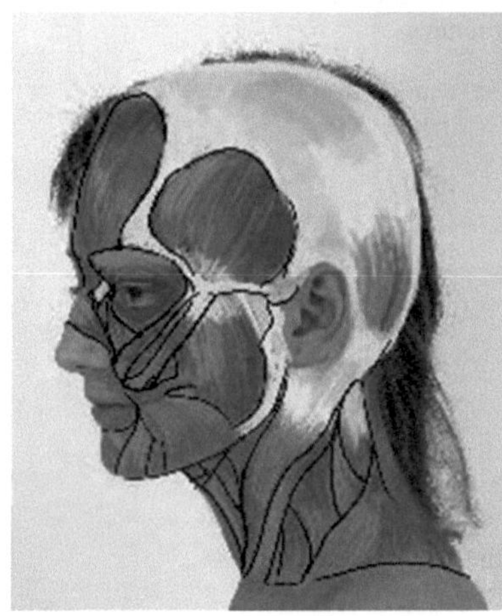

Fig. 3.2 Underlying structure of the head and neck

The next step in developing this understanding of facial structure is to continue the draw-over process. In Figs. 3.3, 3.4, 3.5, and 3.6, another student begins following the various surface areas of the face while referencing medical illustration imagery.

By going through the exercises and manually drawing in the lines, the student begins to develop a feel for the shapes, the forms, and the outlines of the structures that lie underneath the skin and contribute to the shapes, the lines, and the highlights on the surface that represent this structure.

Students are then asked to use these underlying forms to develop a strategy for their edge flow. It is important to understand that at this point, the ultimate modeling method, whether NURBS, polygons, or subdivision Surfaces, is unimportant; the placement of edges for the determination of shapes is the same throughout.

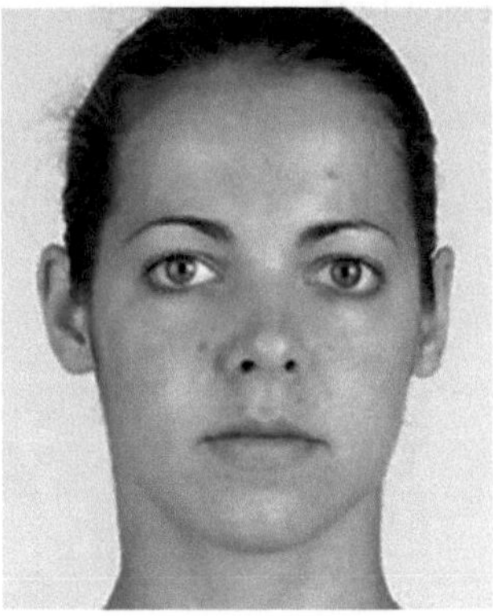

Fig. 3.3 Surface of the face, head and neck, showing the underlying structures

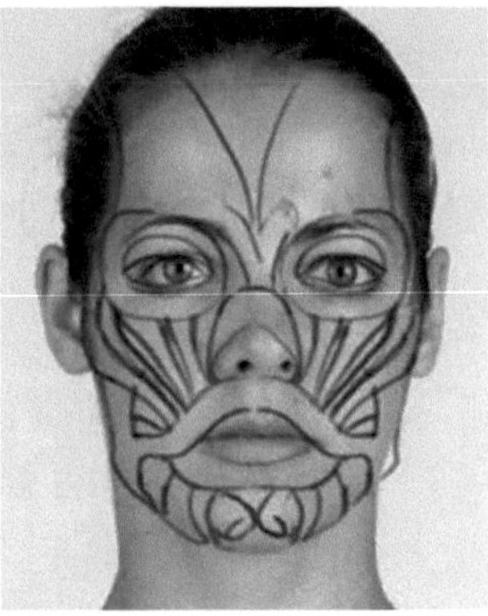

Fig. 3.4 Student overlay illustration of underlying structures

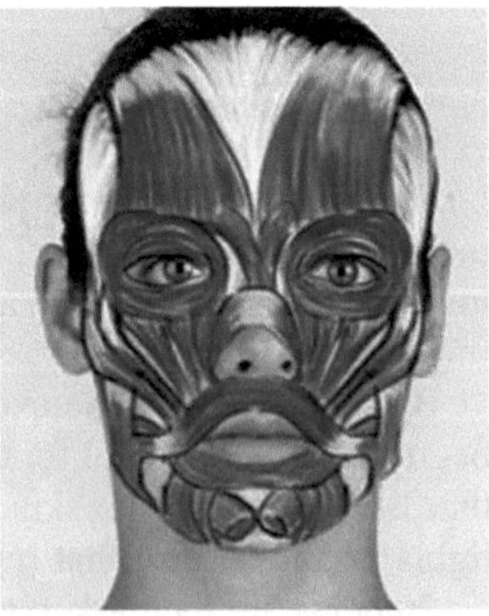

Fig. 3.5 Student overlay illustration showing underlying bone, muscle, and structure

Using printed images and tracing paper or digital images and layers in an imaging program like Photoshop, and with their previous drawings as reference, students lay out their edges.

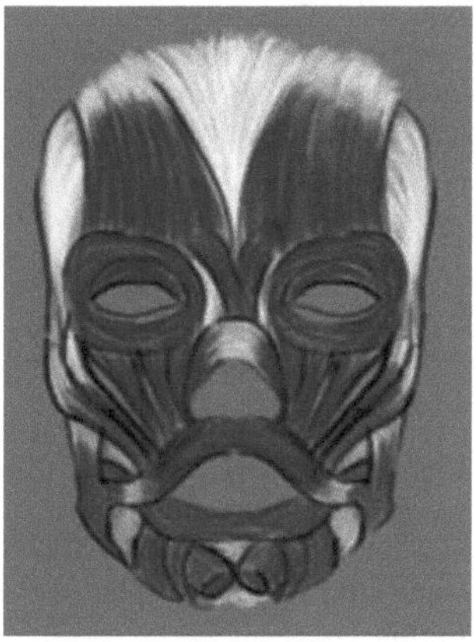

Fig. 3.6 Student illustration of facial bone and muscle

At this point, it makes no difference if the resulting edge flow is targeting a final output in NURBS, polygons, or subdivision surfaces, the goal is always to put the edges where they will perform optimally in terms of defining and deforming the surfaces in the most accurate and efficient way possible.

3.3 NURBS Patch Modeling

NURBS "patch modeling" is a highly technical process, with emphasis placed on mathematical accuracy and respect for the formula-driven creation of curves and surfaces. The ability to sculpt accurately with NURBS is severely hampered by these technical constraints; students learning to model must develop a level of mastery of the workflow before applying it to any creative endeavor.

NURBS surfaces are defined in one of the two ways, either "uniform parameterization" or "chord-length parameterization." These terms refer to the underlying mathematics of the surfaces.

"Uniform" parameterization refers to the numeric identification of each isoparm in the *U* or *V* direction of the surface. Beginning at the origin of the surface, the isoparms in *U* and *V* are numbered sequentially, beginning with zero.

Uniform parameterization can either be expressed by the total number of isoparms (which define the number of spans) using sequential integers (0, 1, 2, 3 …) or by using evenly spaced decimals within a 0–1 numeric system. In either case, the position of any given span relative to the world-space measurement of the surface is disregarded (Figs. 3.7, 3.8).

In contrast, chord-length parameterization does take into account the world-space measurement of the entire surface when assigning numeric values to the individual isoparms. The first isoparm is given a numeric value of 0, and the numeric value of the last isoparm reflects the entire length of the surface in world

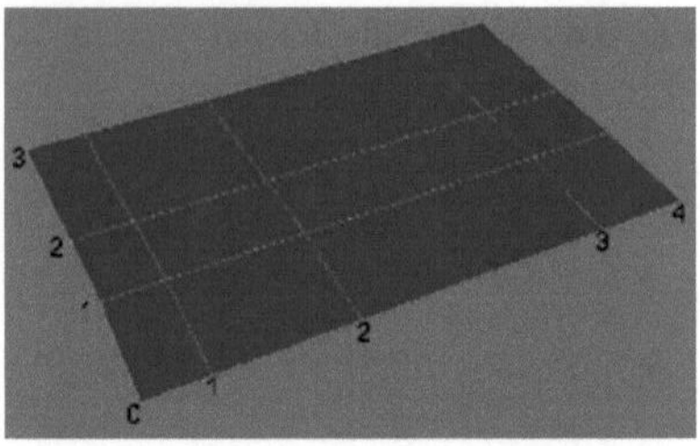

Fig. 3.7 Numeric values assigned to isoparms in a uniform NURBS surface, with values arranged as "0–#Spans"

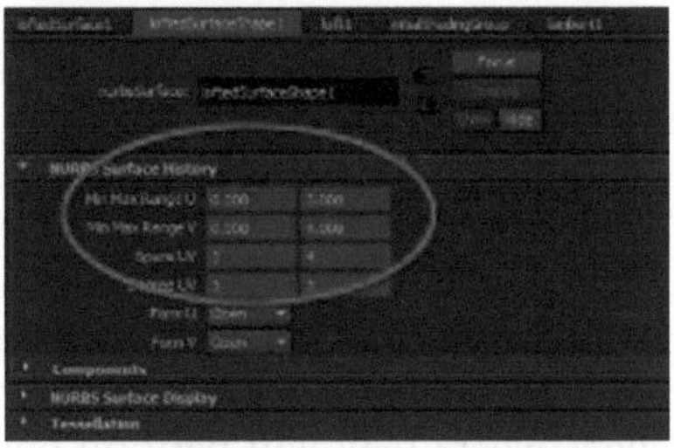

Fig. 3.8 Numeric values in the *U* and *V* directions are also reflected as min/max values in Maya's Attribute Editor

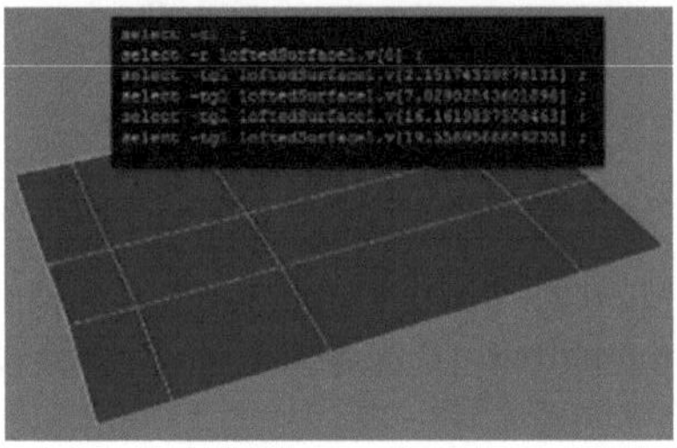

Fig. 3.9 The isoparms in a chord-length surface are assigned scalar values, which are ratios that represent the position of the isoparm relative

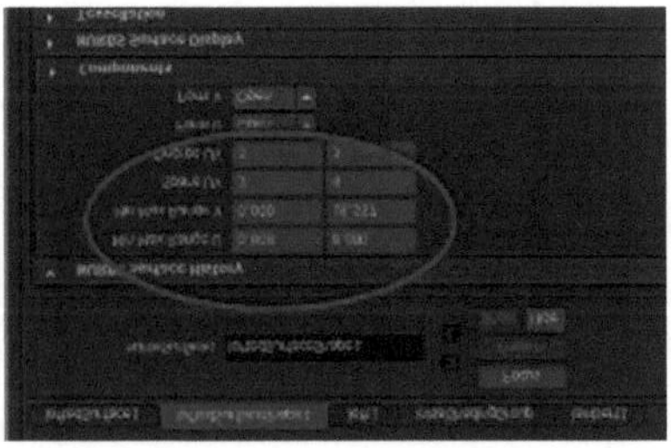

Fig. 3.10 Just as is the case with uniform parameterization, numeric values in the *U* and *V* directions are also reflected as min/max values in Maya's Attribute Editor

space. The value assigned to any given interior isoparm is a ratio of that position relative to the entire surface length in the given direction (Figs. 3.9, 3.10).

Whether uniform or chord length, NURBS patches must be "square"; in other words, the number of patches at the beginning of the *U* or *V* direction must be the same at the end; no holes or detachments are allowed. Single NURBS patches cannot fork in two directions (Figs. 3.11, 3.12).

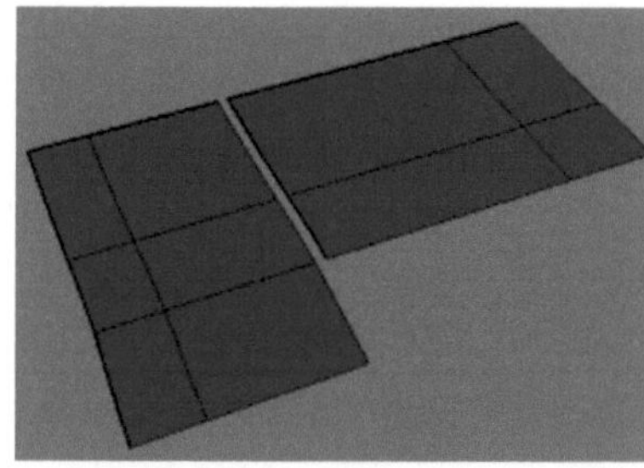

Fig. 3.11 This configuration is impossible in a single NURBS surface

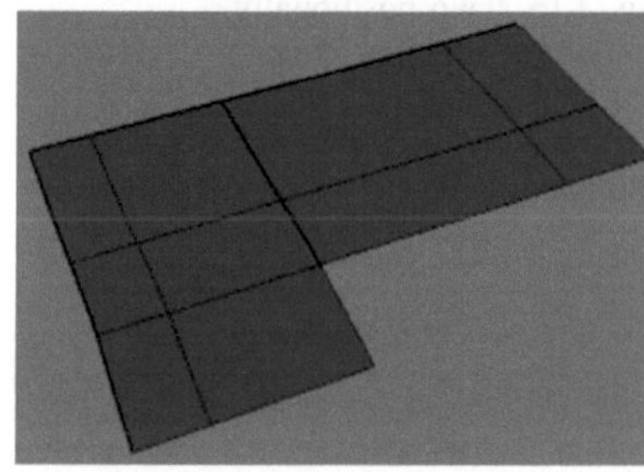

Fig. 3.12 What you actually see is two surfaces that are touching

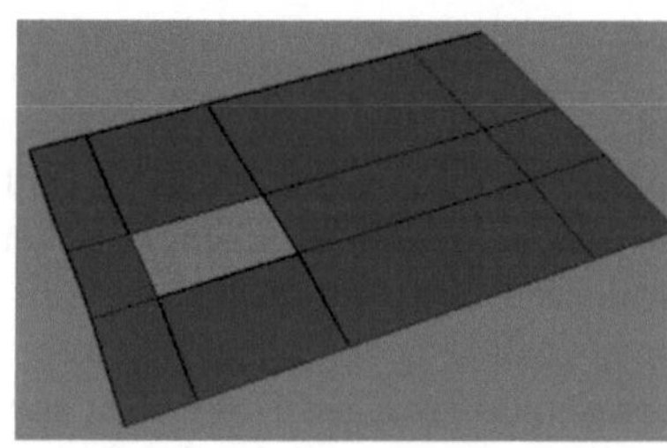

Fig. 3.13 A hole like this is also impossible in a single NURBS surface

It is important to note that when creating complex surfaces, holes may be absolutely necessary, and the edge flow of models must change direction frequently—which presents an inherent technical problem for the NURBS modeler. In order to solve this problem, surfaces must be broken up into individual square patches and stitched together, maintaining continuity between adjacent surfaces (Figs. 3.13, 3.14).

"Surface continuity" refers to the way in which one surface connects to an adjacent surface. In the field of engineering and industrial design, multiple options for surface continuity with NURBS models can be achieved. Note that the degrees of surface continuity mentioned below are sequential; the set of rules defining any level of continuity contains all the rules of the level beneath it, with additional factors added.

G-0, or "positional continuity," requires only that two surfaces share an edge, that they are tangent to each other. No consideration is given to the quality of that edge; it is generally assumed to be a hard corner.

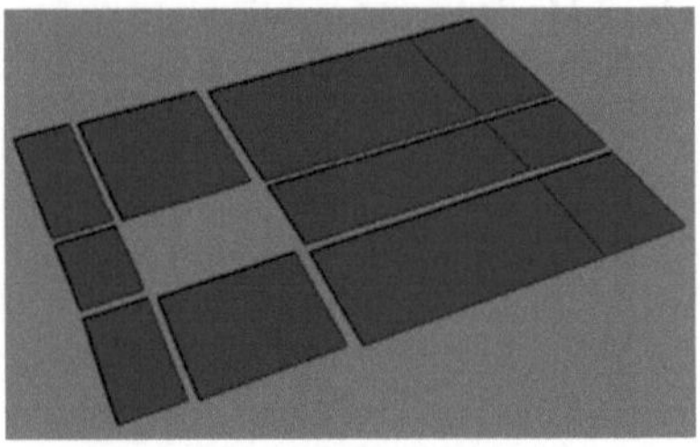

Fig. 3.14 What you actually see is eight surfaces that are touching

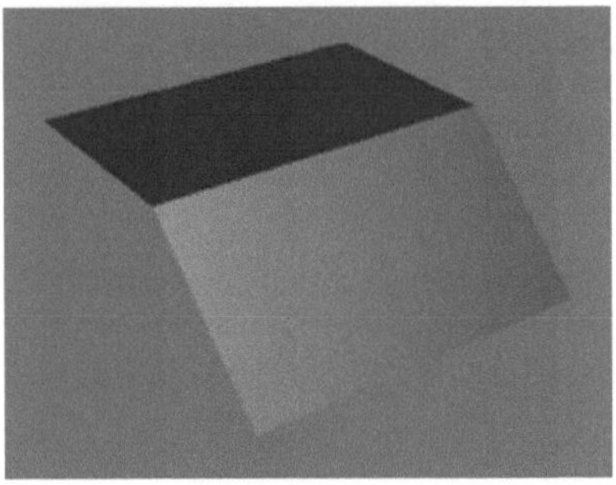

Fig. 3.15 Two positionally continuous surfaces. Note that their shared edge is marked by a hard break in illumination

G-1, or "tangent continuity," means that two surfaces share not only an edge, but a tangent line at the point of positional continuity, as measured at the surface normal. If the normals are also both perpendicular to the same tangent line, they are also coincident. Coincident end-normals that reflect positional continuity produce a transition from one surface to the next with no visual breaks. These breaks would generally be apparent when light shining on the surface is calculated, represented by a hard line in a specular highlight or diffuse attribute of the surface's response to light. Deviation from absolute coincident can be acceptable based on predetermined tolerance factors and is typically calculated in nonzero angular measurement between the two normals in question.

For organic modelers, character modelers, or digital sculptors, tangent continuity is typically the ideal, which will be discussed in more detail later in this chapter (Figs. 3.15, 3.16).

G-2, or "curvature continuity" is of more importance to engineers. In addition to a shared tangent at the point of positional continuity, G-2 continuity requires equivalent rates of curvature for both surfaces as additional spans with increasing parametric distance from the point of transition are considered.

For the purpose of engineering, industrial design, and manufacturing complex surfaces, additional levels of surface continuity are possible; however, for the purpose of this chapter, we will stay focused on G-1, or tangent continuity.

From a practical, rather than theoretical standpoint, the test for tangent continuity is relatively simple, and this kind of continuity can be achieved in a number of ways.

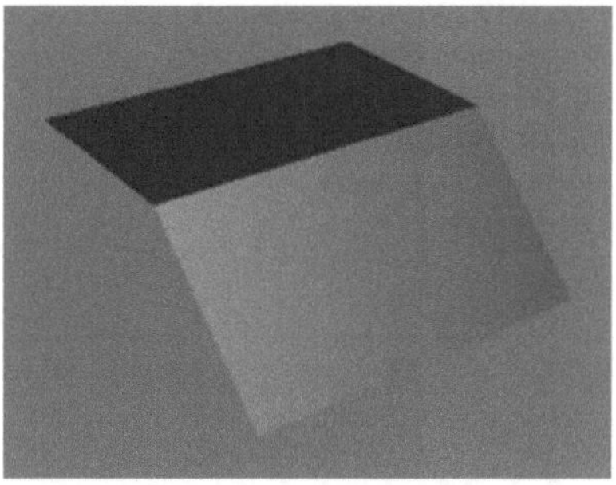

Fig. 3.16 Two tangent-continuous surfaces. There is no break in the illumination where the two surfaces join

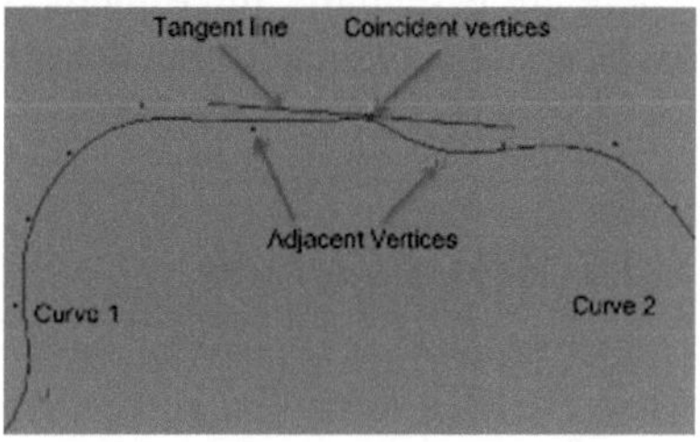

Fig. 3.17 Because the end vertices of curve 1 and curve 2 are coincident, these two curves are positionally continuous. However, the tangent line cannot intersect the coincident vertices and both adjacent vertices, so no tangent continuity

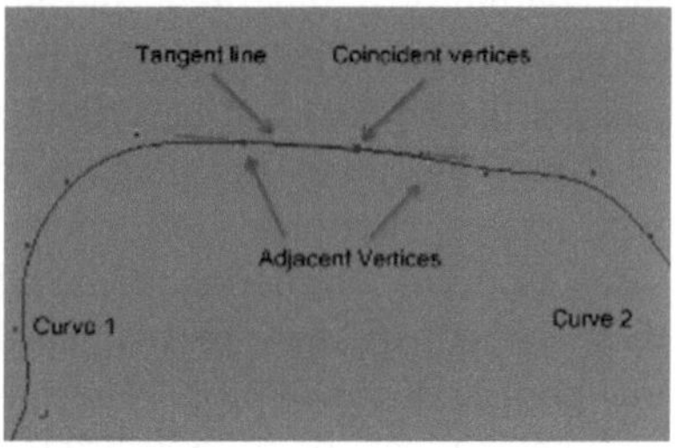

Fig. 3.18 The adjacent vertices have been moved so they now intersect the same tangent line that passes through the coincident vertices. These curves are tangent continuous

For two curves to be tangent continuous, the end vertices (CVs) must be coincidental (positional continuity). If a straight line can be drawn from the vertex adjacent to the end on one curve to the vertex adjacent to the end on the other, and this line intersects the coincidental end vertices, then the two curves are tangent continuous (Figs. 3.17, 3.18).

By extension, the same test is given to surfaces. While it is not a mathematical requirement that adjacent surfaces have the same number of spans, it is a practical one. For purposes of determining and achieving tangent continuity between two surfaces, the coincident vertices at the end of each span and their adjacent vertices must pass the same test, above (Fig. 3.19).

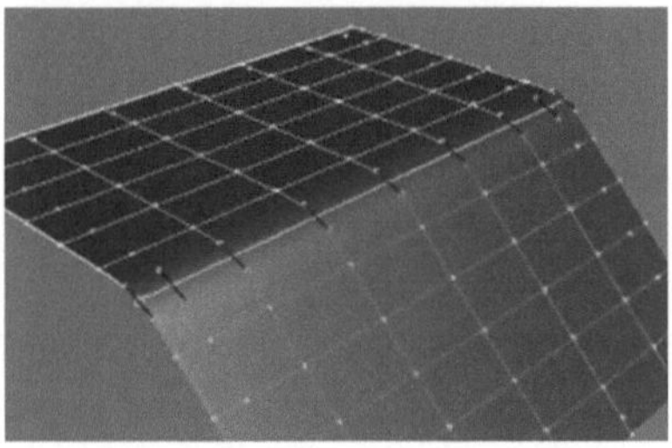

Fig. 3.19 Vertices adjacent to the coincident vertices are on the same tangent line (*blue lines*); therefore, these two surfaces are tangent continuous

3.3.1 Methodologies

There are essentially three methods of achieving tangent continuity between two surfaces: attach/detach, manual alignment, and automatic stitching.

3.3.2 Attach/Detach

Most NURBS modeling software packages have an "attach" function, where two surfaces can be made into one. A common option within this function is "blend," which creates one continuous, smooth surface. When this surface is then detached, the continuity is inherited, and an inspection of the adjacent vertices along the common edge reveals the placement of "tangent vertices," with each set of four passing the tangent test described above.

A prerequisite to this attach is that the surfaces have the same parameterization type (either "uniform" or "chord length") and, if they are uniform, follow the same isoparm numbering convention ("0–1" or "0–#of spans"). While any of these will satisfy these prerequisites, it is generally accepted that "uniform, 0–1" is the preferred convention. Therefore, if two adjacent surfaces are of different parameter types, or follow different numerical conventions, they must be rebuilt prior to the attaching.

Further, NURBS patches should be aligned in the same direction—with their *U* and *V* directions parallel and their normals facing the same direction. The beginning of one surface (Isoparm #0) should attach to the end of the other surface (Isoparm #1). If an inspection of these surface parameters reveals non-alignment, the surfaces must be reversed in either or both directions, and/or the directions must be swapped.

Finally, when two surfaces are attached, the numerical values assigned to the isoparms will change. When they are detached, these values do not revert, so the surface must again be rebuilt.

As complex as it may appear here, the attach/detach method of surface continuity is the simplest process to follow when creating organic models using NURBS. The workflow becomes rote: *rebuild, reverse, attach, detach, rebuild,* and *delete history* (Figs. 3.20, 3.21).

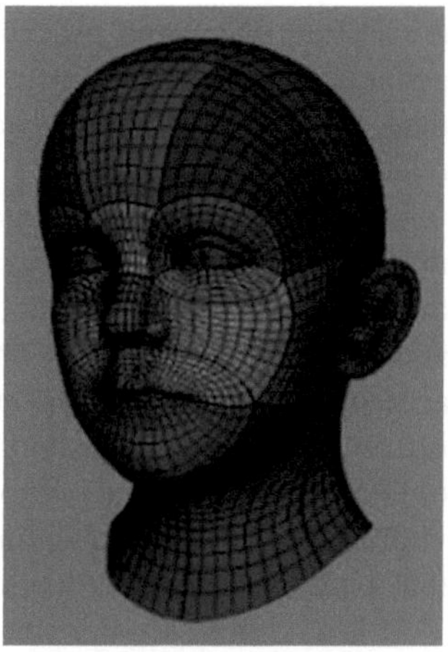

Fig. 3.20 A complex NURBS model must be pieced together with square patches that meet all of the conditions described in this chapter

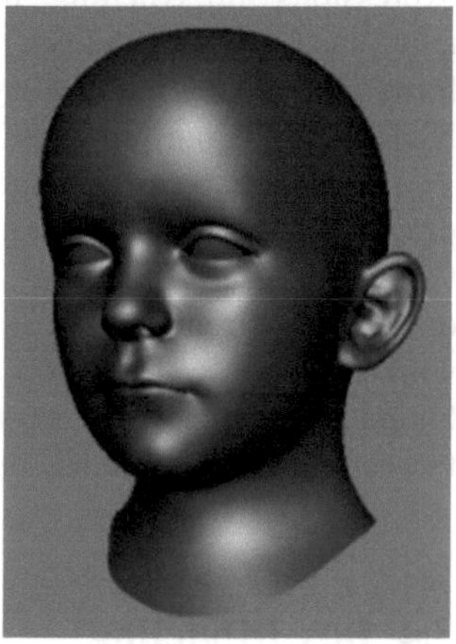

Fig. 3.21 If any surface in the model maintains tangent continuity with respect to its adjacent surfaces, the result will be a smooth model, with no breaks in surface highlight, giving the appearance of a single

Once a model has been pieced together using the above methods, after a fair amount of cleanup and reworking, inconsistencies, pinch-points, and non-tangent-continuous surfaces will be discovered. Fortunately, these areas are not too prevalent and can usually be fixed using manual methods of achieving continuity. Usually, these methods involve drawing a straight line between two vertices, extending that line and snapping the remaining vertices into place using a curve snap or point snap function. Several users have written simple scripts to simplify this process.

The other method of achieving tangency is automatic and software-specific. Usually, this is a tool called "Align Surfaces" or "Stitch" or "Global Stitch." The tool is simply an automated way of performing all the manual tests outlined above (positional continuity, parametric consistency and alignment, and normal-angle coincidence) across multiple surfaces, making a series of corrections based on preset tolerances and procedural algorithms. From a practical point of view, it is generally a bad idea to depend too heavily on these automated methods initially; they are best used in small areas or in areas where corrections are few and relatively minor.

With all the technical requirements and complexities inherent in NURBS patch modeling described above, it is important to note that we have not yet addressed the most important aspects of organic modeling in general—which are *edge flow* and *sculptural aesthetics*. These important considerations must also be taken into account during the process of *rebuild, reverse, attach, detach, rebuild,* and *delete history*. It is for this reason that NURBS modeling (as a final output) has fallen out of favor among studios. A solid mix of technical ability and aesthetic expression is vital to this style of modeling and is contrary to the compartmentalization model that photogrammetry allows.

It is also difficult and computationally expensive to maintain tangent continuity among surfaces that deform during animation. With each deformation (facial expressions, lip-synching, muscle or other organic deformation), tangent continuity must be recalculated and adjusted—typically through adherence to some form of construction history algorithm. This, too, is a major drawback to the NURBS modeling pipeline.

This is not to say that there is not a place for NURBS modeling, either in the process of teaching modeling or within a production pipeline. NURBS tools are very robust and, sometimes, are the preferred method for achieving certain shapes or volumes. From an educational perspective, the discipline required of a NURBS work-flow helps to develop a naturally clean, efficient approach for students, and builds an inherent tendency to seek accurate and efficient edge flow that is vital to the aesthetics of a model, but not necessarily required for its technical success.

3.4 Box Modeling

Box modeling is the method that is most typically used and taught as the traditional method for polygonal organic modeling. As the name implies, box modeling begins with a simple primitive shape, typically a cube, although in many cases, a model might begin with a sphere or cylinder.

The goal with this kind of model is to define the size and proportion of the figure while exercising restraint in the process of adding edges until they are needed. For this reason, an understanding of the major shapes that make up the human form is vital and plays a big part in the students' decisions as to where to introduce new edges.

In Fig. 3.22, the model was started with a cube that was scaled to the proportions of a human torso. In Fig. 3.23, the first edge introduced is the waistline, which helps to differentiate the shape of the upper body (chest, rib cage) from the lower (pelvic girdle). The centerline is introduced in Fig. 3.24 with two adjacent edges which will differentiate the groin from the legs, which will be extruded or

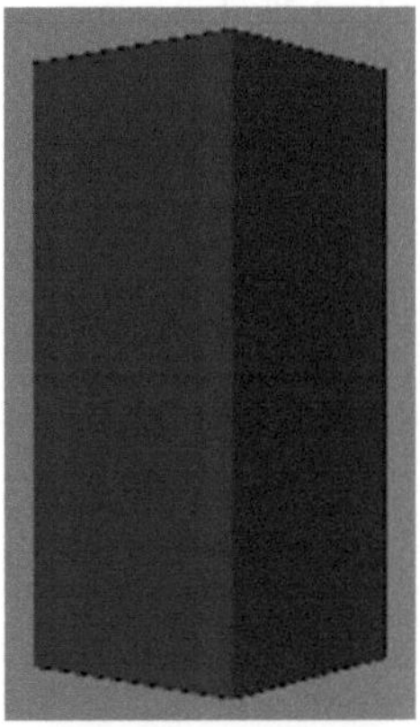

Fig. 3.22 A simple cube is the beginning of a human torso

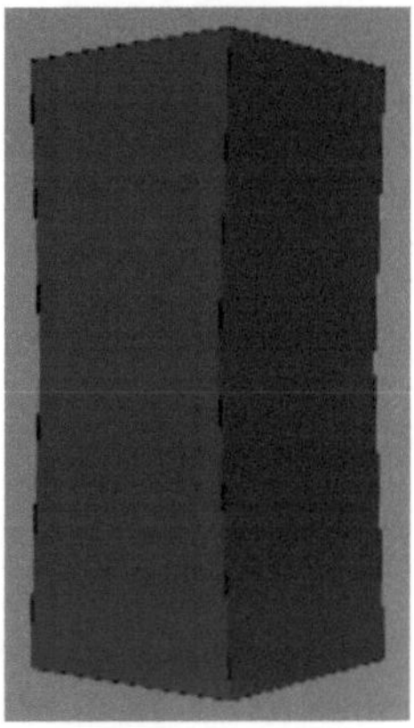

Fig. 3.23 Major character lines are inserted first

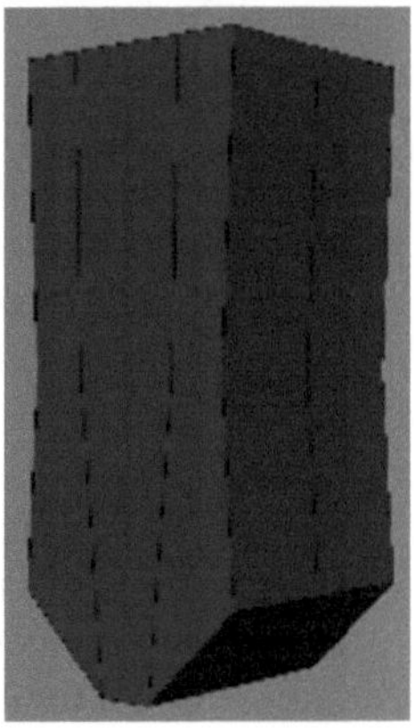

Fig. 3.24 Additional lines are inserted in order to begin rough-in of general shape

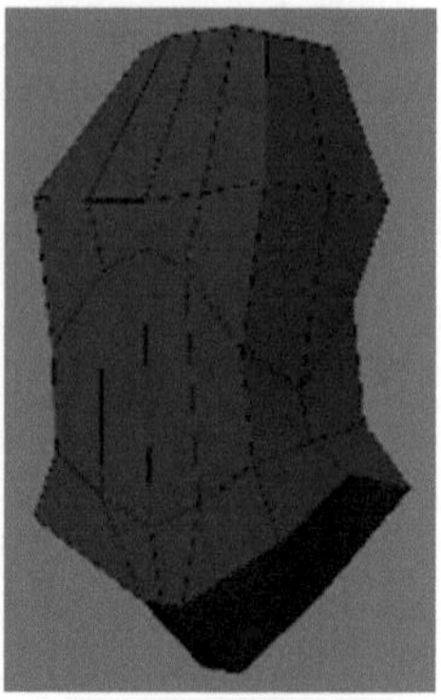

Fig. 3.25 Lines are added to define hips, rib-cage, shoulder area

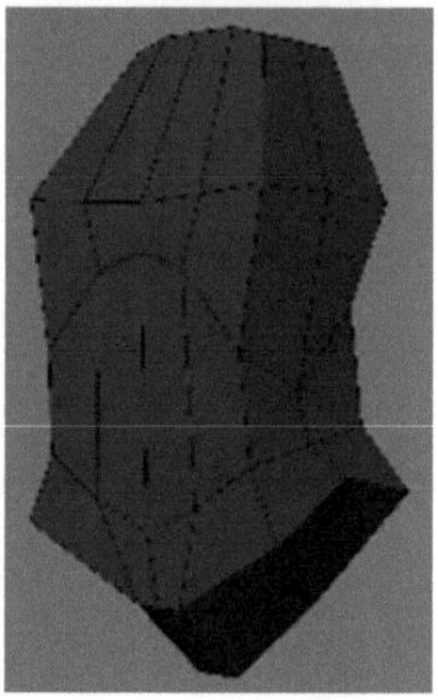

Fig. 3.26 Lines are averaged to achieve more roundness

attached later. Vertices are moved up along the outer pelvic area to define the dish shape of the pelvis and to simulate the natural lines of the abdomen resting within.

With the placement of each edge, the model is evaluated as to its likeness to the reference material, including the anatomical references and draw-overs previously executed. In every case, the attempt is made to refine the shape as much as possible without the addition of new edges.

As the model progresses, the existing edges are manipulated into positions where they can define specific structures beneath the skin. In Fig. 3.25, the waistline is moved up to define the lower part of the rib cage, while additional edges are added to create the pelvic structure and the chest.

In Fig. 3.26, no additional edges have been added but the existing vertices have been averaged, or softened, to relax the cube shape and make it appear more cylindrical.

In Fig. 3.27, additional edges were added to the chest, which are then used as extrusion points to create the deltoid and the upper arm.

In this way, shapes are visualized and edges are placed to achieve these shapes. Attention is continually paid to the original draw-overs that identify the larger shapes, than the smaller ones—from planes to masses to individual muscles and bones (Figs. 3.28, 3.29, 3.30, 3.31).

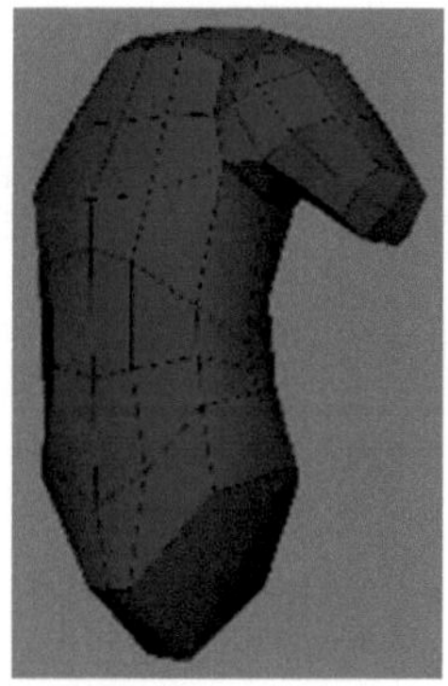

Fig. 3.27 A reasonably accurate shape can be attained with very few lines

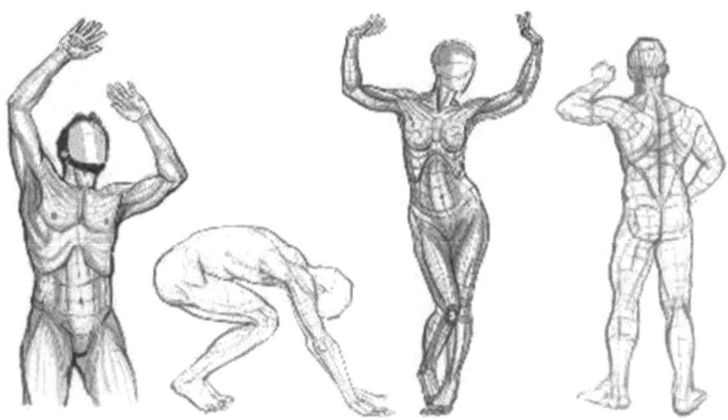

Fig. 3.28 Student draw-over illustrations, showing major "landing zones" for polygon edge-flow strategy

As the figure begins to take shape, the edges begin to create rows of individual polygons, which can then be shaped into the forms observed and drawn in the earlier exercises.

3.4.1 Langer's Lines

Another consideration is the natural flow of tension lines along the skin, known as "Langer's iines." Karl Langer was a nineteenth-century surgeon who produced a series of diagrams depicting lines of cleavage in the skin. Anatomically, these lines correspond to the natural orientation of collagen fibers in the dermis and are an important consideration for surgeons in deciding the placement and orientation of

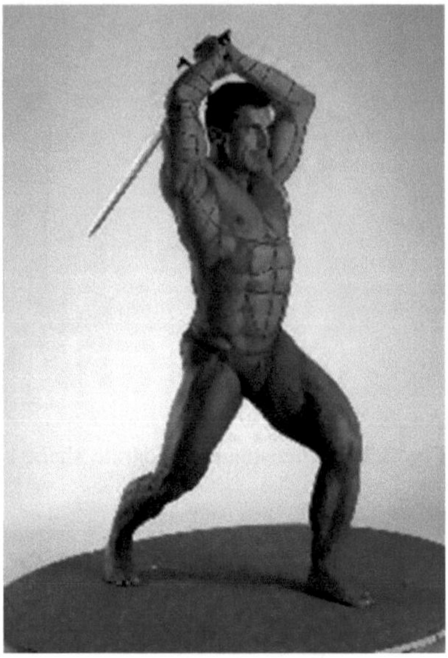

Fig. 3.29 Student draw-over illustrations, showing major "landing zones" for polygon edge-flow strategy

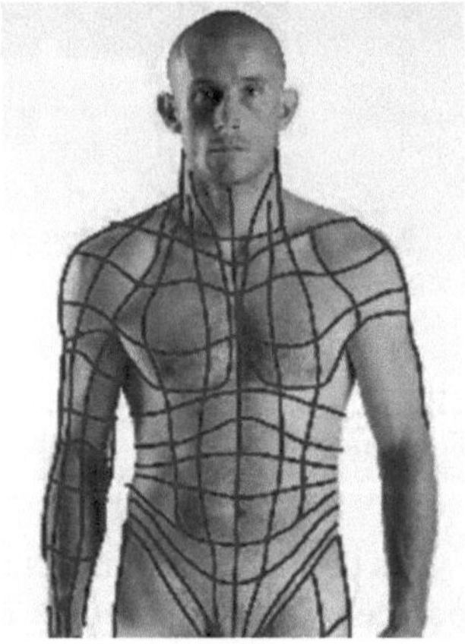

Fig. 3.30 Additional strategy for polygon layout

surgical incisions. If an incision follows Langer's lines, the scar will align with the natural tension lines of the skin and will heal in a more camouflaged state.

> Collagen is what dictates the shape and direction of wrinkles in the skin—those formed by facial expressions, body position and aging. Langer, Karl (1861)

For artists, these lines are also important. When determining proper edge placement for a model regardless of the method being used, if the edge flow describes the underlying shapes and creates a matrix for realistic deformation, more realism can be achieved (Figs. 3.32, 3.33).

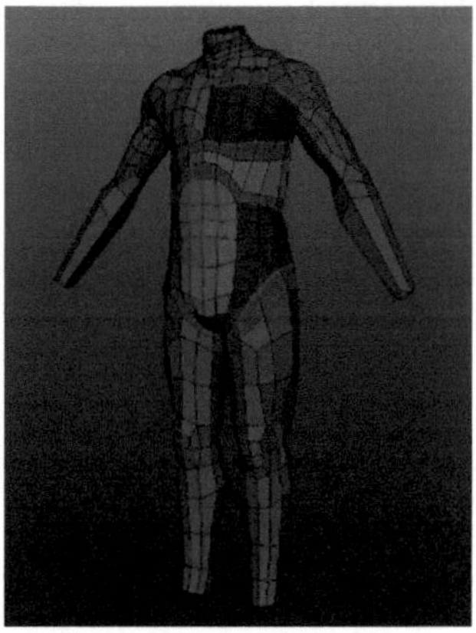

Fig. 3.31 Finished layout, following the strategies developed in the draw-overs

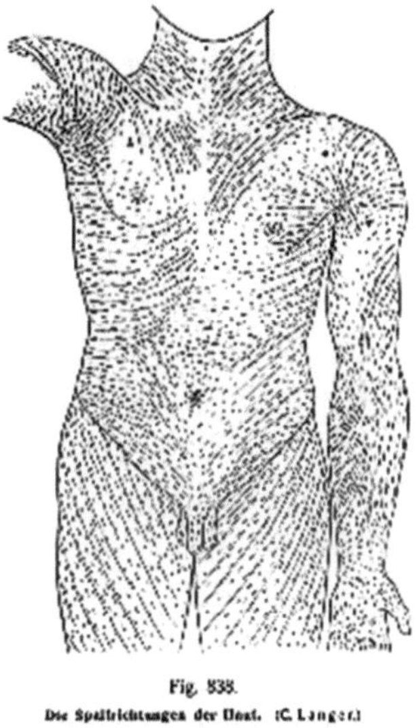

Fig. 3.32 Langer's Lines of the torso and upper appendages

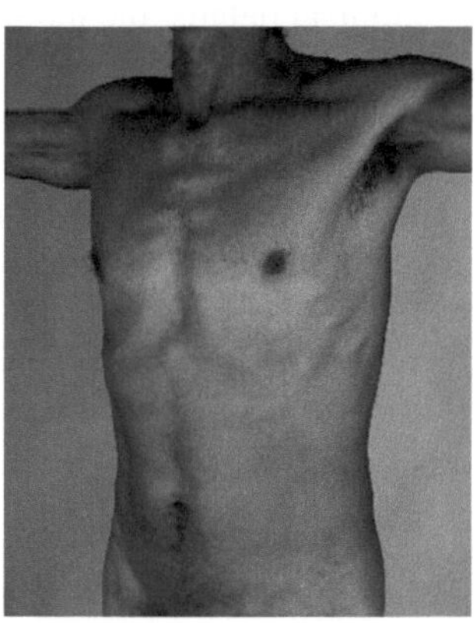

Fig. 3.33 Langer's lines revealed through the flow-lines of the skin

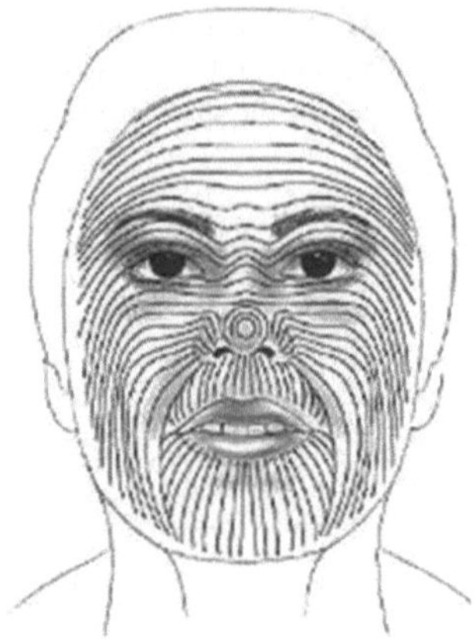

Fig. 3.34 Langer's lines of the face

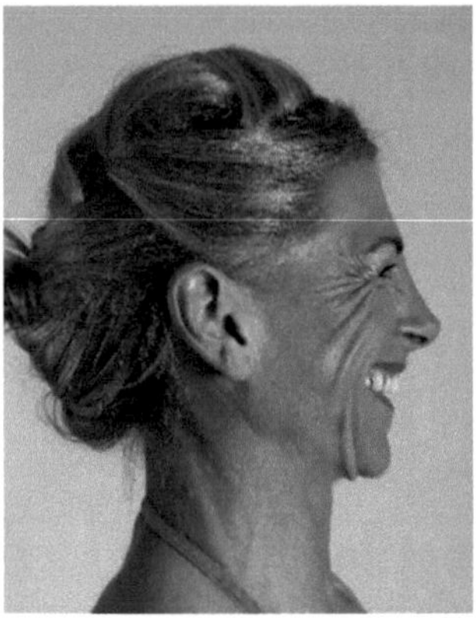

Fig. 3.35 Facial wrinkles showing Langer's lines

Whether designing body or a face, the artist must take into consideration the structure underneath the skin, and the structure of the skin itself when determining the proper placement of edges, lines, and polygons. By faithful attention to these anatomical details, the modeler can achieve a higher degree of realism (Figs. 3.34, 3.35).

All of these details present a very practical problem to the student who is first learning how to create an organic model. For many, as described in previous chapters, the technical and aesthetic challenges are often mutually exclusive. As educators, we are frequently faced with the dilemma that "you have to know everything before you can learn anything." The process of organic modeling, therefore, is frequently one of trial and error—discarding early versions and starting over as new concepts are learned on both sides.

3.5 Normal Mapping, Displacement Mapping, and UV Layout

In the world of 3D graphics, the notion of "location" can be variable and must be clearly defined within the context of the specific tasks we are undertaking.

Picture, for the moment, the location of your home. Within a certain context, you can regard this location as a constant—you always know where your home is, you can find it on a map and in the world. It never moves. However, your home is on the planet Earth, which revolves on its own axis, orbits the sun, and is part of a solar system that traverses the universe. In a broader system of coordinates, those that define the universe for example, the location of your home is constantly changing.

This is an example of the difference between "world space" and "local space." In a 3D modeling system, we typically define the Cartesian grid system as the universe where everything takes place. Position in this universe is constant; the *X*, *Y*, and *Z* axes define our world. We can keep track of the position of our objects and the variations of their surfaces by defining the *X*, *Y*, and *Z* positions of the individual vertices in our calibrated universe. This is "world space."

However, we also have a need to define the position of a given point relative to a surface. If I want to place a beauty mark on the cheek of a model of Marilyn Monroe, I want that beauty mark to remain in a constant position relative to the model, regardless of the position of the model in world space. This different, but equally important, location system is commonly known as UV space. "U" and "V" are names given to a set of surface-based coordinates by which we are able to find our way around the surface of a model.

This kind of local space definition has been used for centuries, in the field of cartography. In the image of the roadmap in (Fig. 3.36), we use the row of letters at the top or bottom and the column of numbers at the sides to find the location of the area of the map we want. In this example, the "Catedral" is in the box defined by G-9. Of course, hyper-accurate location information is also available via GPS data, which can provide latitude and longitude information accurate to within a few feet.

In computer graphics, the layout of *U* and *V* space is only slightly more complex than locating points of longitude and latitude. A 3D surface is "unfolded," or laid flat—much like a Mercator or other projection of the globe onto a 2D space. A calibrated, 2D grid is then projected onto this unfolded surface, which is then used to identify coordinate space relative to the surface itself rather than world space.

In Figs. 3.37 and 3.38, the sculpted head at the left has been unfolded and laid flat; the flattened image is shown in Maya's UV texture editor. By using the XY grid overlaid on top of the organic, irregular shape of the unfolded head model, specific points on the surface can be identified for the purpose of applying color, texture, and other specialty "maps" such as normal maps.

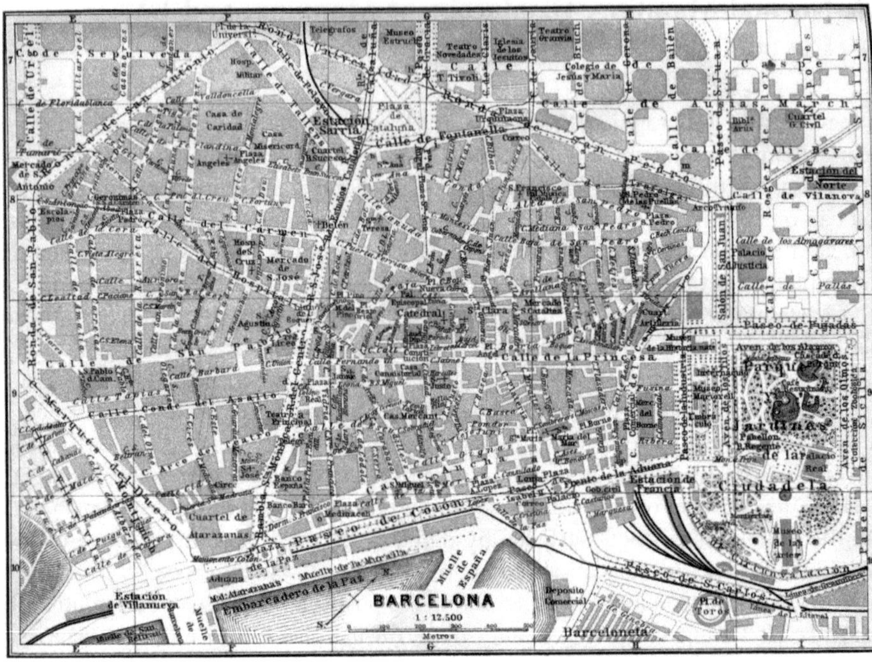

Fig. 3.36 Roadmap

Fig. 3.37 Sculpted head

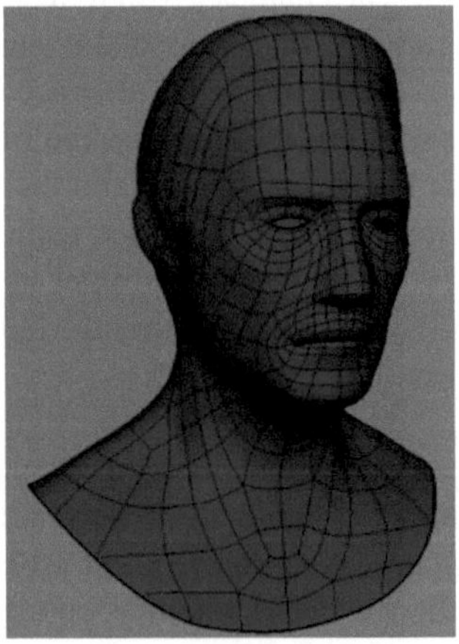

In 3D computer graphics, the term "normal" refers to an imaginary line that is perpendicular to a line tangent to a surface at any given point. The direction of these normal vectors are compared to the vector of the line of sight of a camera or a light, and the resulting angle (or "facing ratio") can be used to calculate any number of dynamic visual relationships. For example, in a Lambertian

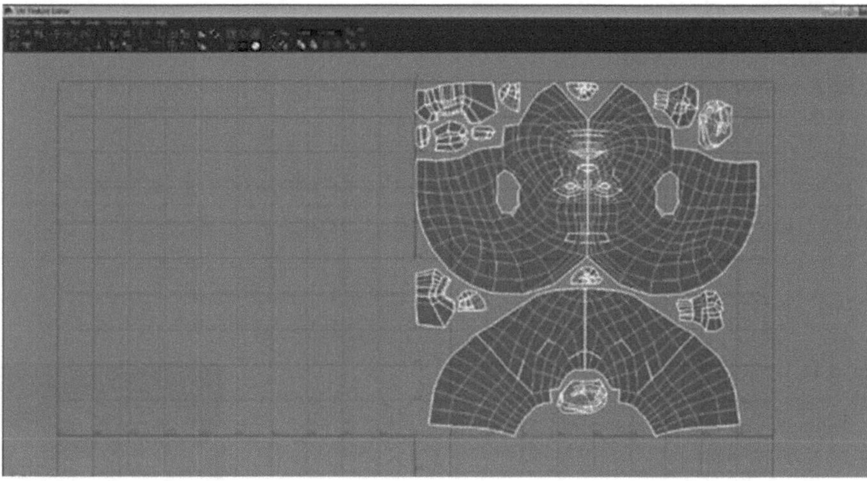

Fig. 3.38 UV "Roadmap" unfolded

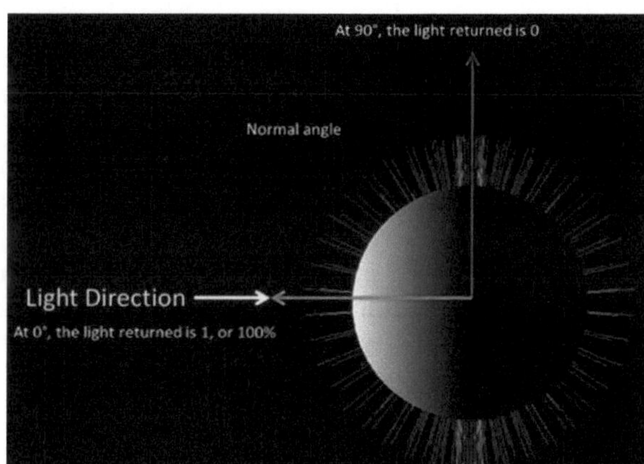

Fig. 3.39 Angle of the light is calculated relative to the surface normal vectors

illumination calculation (illustrated below), the amount of reflected light energy returned by a given surface is equal to the incident energy multiplied by the cosine of θ, where θ is the angle created by the incident light vector and the surface normal, multiplied by a diffuse value determined by the given material. This is known as "Lambert's Cosine Law" and provides a realistic representation of light falloff as the angle of the surface changes relative to the angle of the light.

In Fig. 3.39 we can see that when the angle formed by the light and the normal vector is zero (the light is pointing directly at the surface), the reflectance value is 1 ($\cos 0 = 1$); and when the angle formed is 90, the reflectance value is 0 ($\cos 90 = 0$).

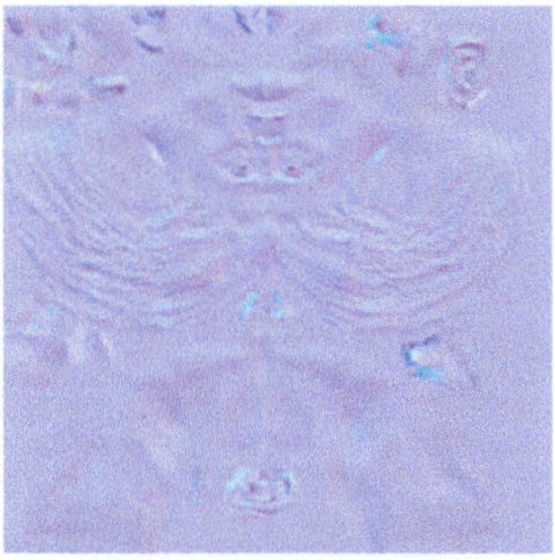

Fig. 3.40 The angle of the light relative to the direction of the surface is represented by RGB color information in a "normal map"

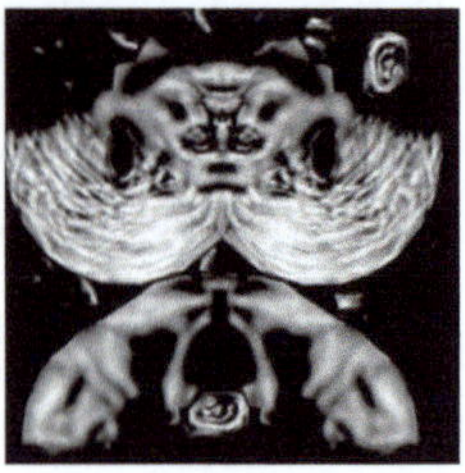

Fig. 3.41 The surface differences between the high-res sculpture and the low res form are represented in grayscale for use as a displacement map

The process of normal mapping uses this Lambertian cosine model to translate the detail of a high-polygon model into a low-polygon model, using the following process:

First, a sculpted model must be topologized correctly, as shown in previous sections of this book. The model is then unfolded, and a UV map is extracted, providing accurate surface coordinates.

The model is then imported into a digital sculpture application such as Autodesk Mudbox or Pixologic ZBrush, where the resolution density can be increased incrementally for more detailed sculpture. As shown earlier, these applications are designed to handle densities of millions of polygons, which provide the capability of sculpting very fine detail such as wrinkles or fine skin texture. The density of the model is increased systematically—with each face divided by a factor of four—so that each level of resolution can "inherit" the same UV texture space as the low-resolution version.

An analysis of the normal vectors is then performed on the high-resolution model; the normal angle of each of the millions of faces is determined and stored as an RGB value in a bitmapped image (Fig. 3.40). This bitmapped image is then applied to the low-resolution image and used as a guide for perturbing the normals so that they are systematically "misread" by the camera. Since the normal vectors are a key component for the Lambertian diffuse illumination model, the result is a highly detailed embossing pattern that exactly replicates the look of the higher resolution geometry.

At the same time, another kind of map is derived, called a "displacement map" (Fig. 3.41). This is a grayscale image that acts in a similar fashion to the normal

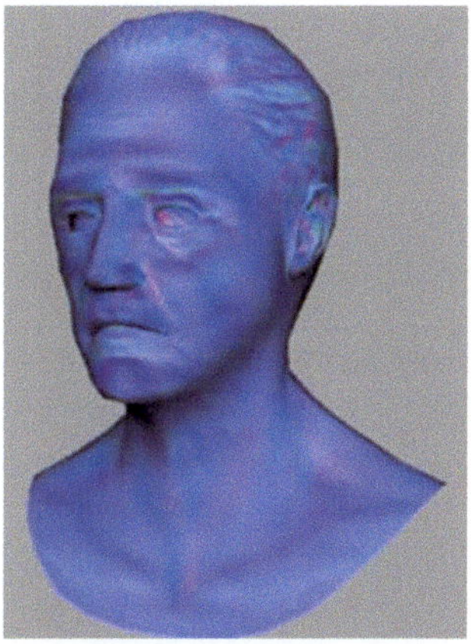

Fig. 3.42 The RGB information contained in the normal map is interpreted as an embossing layout, to show surface detail

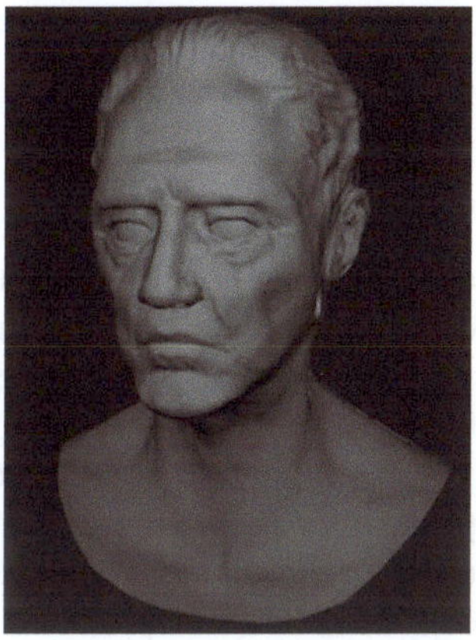

Fig. 3.43 Low resolution sculpture appears to be very highly detailed with the application of the normal maps and the displacement maps

map; however, with this kind of map, the alpha channel of the image will physically displace the low-resolution geometry to approximate the changes in surface contours represented by the high-resolution image.

The combination of the displacement map, which changes the actual shape of the geometry, and the normal map, which acts as an "embossing" filter to represent the finer details such as wrinkles, lines, and pockmarks, allows for an extremely high level of detail on a relatively low-resolution model.

Figure 3.42 shows a screen capture of the low-resolution image with the normal map applied as color, for display purposes only. Notice how the patterns of color line up with the finer detail lines of the finished sculpture (Fig. 3.43).

The accepted workflow is thus "Sculpt"—Retopologize—Derive UV Map—"Up-Res"—Sculpt—Derive Normal Map—Derive Displacement Map—"Down-res"—Apply Maps—Render.

It should be noted also that the use of Normal Mapping only works with polygonal geometry, not with NURBS. This is one major reason that NURBS

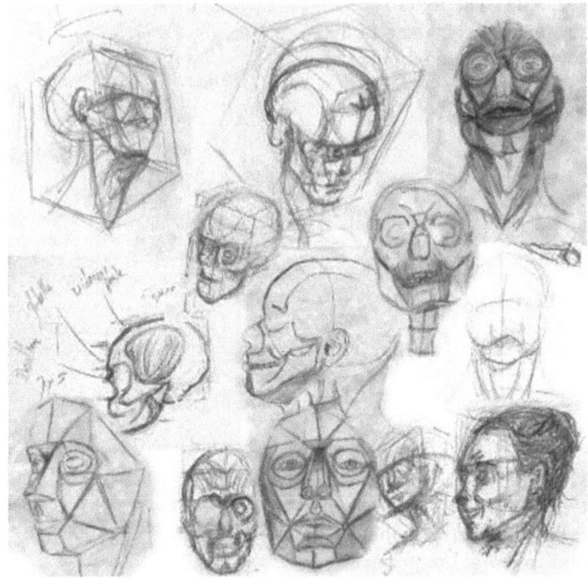

Fig. 3.44 Student sketchbook

modeling has fallen out of favor among artists and studios. The level of sculptural detail that can be achieved using the normal/displacement mapping workflow has made polygon modeling the preferred choice, particularly in the context of the heavy technical needs of NURBS.

The important point here, which relates to the pedagogical models we are presenting, is that the technical processes have become less intensive for the end user. The retopology and UV layout are relatively simple procedures and the derivation and interpretation of the normal maps are done entirely by the software, which frees the end user to focus most of his energy on the aesthetic aspects of his work.

3.6 Subtractive Modeling Workflow

As outlined earlier, students will assemble an array of images of their subject for use as reference images. Using 2D imaging software such as Photoshop, or with a pad of tracing paper, the student will identify the underlying bone structure, muscle structure, shapes, and proportions. As with any traditional sculpture, a number of rough sketches will be made to show these forms from various angles. Figure 3.44 shows a page from a student sketchbook, prior to beginning on the sculpture project.

Using a voxel-sculpting software such as Pixelogic's ZBrush or Autodesk Mudbox, the student begins to rough out the shape of the sculpture. In some cases, the student will begin with a default primitive shape such as a sphere; in other cases, they can begin with a rough shape that is part of a collection of "starter models" provided by the software manufacturer. Using interactive "push/pull/drag/carve" tools, the starter model is roughed into shape by the user. Although this can generally be done with a mouse, many users prefer a pen-based tablet or Wacom Cintiq tablet.

Figure 3.45 is an example of a default head mesh that comes with Autodesk "Mudbox" software. The next two images are the beginning placement of key landing zones in the sculpture: the defined jawline, cheekbones, brow ridge, and sternocleidomastoid tendons. With very little technical instruction or software training, an artist can achieve a result like this within a very short period of time (Fig. 3.46).

Note the difference in wireframe density between Figs. 3.45 and 3.47. The default head mesh is 2,002 polygons, which is dense enough to achieve a rough shape. As more detail is needed, the density can be increased. Each level of increase cuts each face in half vertically and horizontally, for a 4:1 ratio. Hence, Fig. 3.47, which represents one level increase in density, has a count of 8,008 quadrangles. There are clearly areas where this density is needed, and others where it is not.

Figure 3.48 is a close-up of a higher-density model. With over 16,000 polygons, we can see that it is possible to achieve some finer detail, particularly in the folds of the cheek and nostril.

Ideally, we would want to have polygon edges follow these folds for more efficient models and better muscle deformation, as the edges help to define the shape of the creases. Any area where this happens at this stage (as indicated by the blue line superimposed on the image) is purely by accident; more often, the creases will cut across polygon edges randomly, as illustrated by the red lines.

Generally, large muscle masses and gross shapes can be modeled at the lowest density level; as more details such as creases and folds in the skin are added, the artist will increase the density stepwise, each level representing a polygon count that is four times the previous level. The artist can move easily through the levels of density, stepping down to change larger shapes, stepping up to add wrinkles and other details.

In Fig. 3.49, we can see that four levels of density increase, or just over 32,000 polygons, are enough density to get a pretty decent amount of detail. However, in the adjacent image (Fig. 3.50), we can see that much of this density does not follow the natural lines of the face. Once we have achieved this level of complexity, it is time to begin the process of retopology. It is important to identify the point at which to stop this phase of sculpture, as the process of retopology can lead to a significant loss of fine detail. Ultimately, we look to achieve no more detail that is illustrated above; we can continue our sculpture on the retopologized model more efficiently and with far less density.

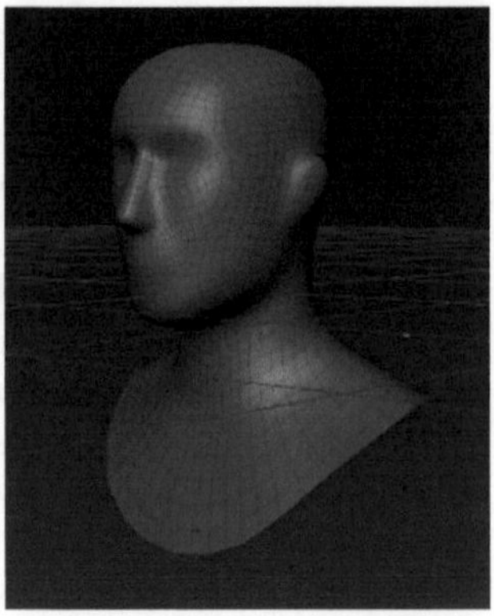

Fig. 3.45 Default head mesh

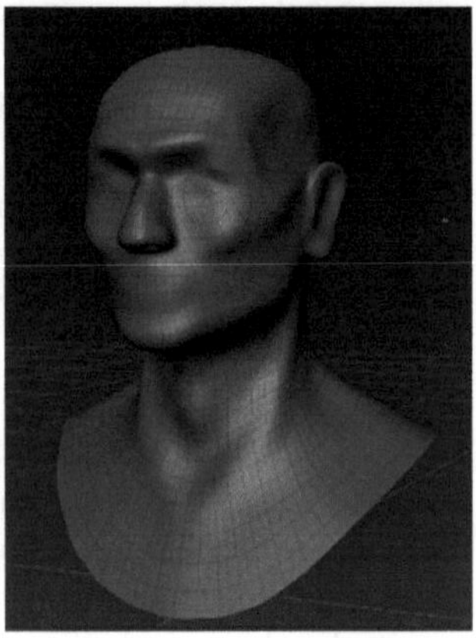

Fig. 3.46 Sculpture in-progress, showing high-density mesh

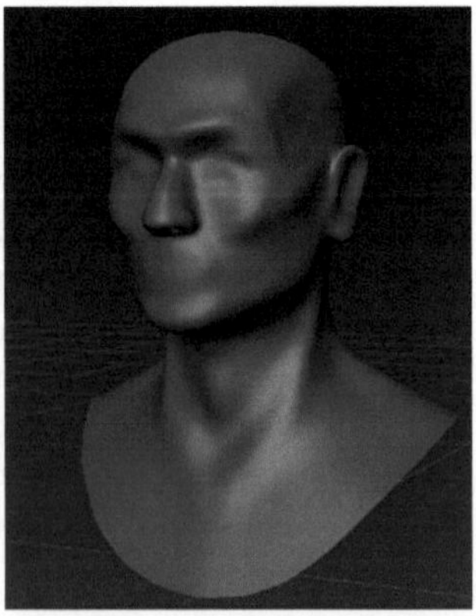

Fig. 3.47 Sculpture in-progress showing no mesh, focused on underlying facial structure

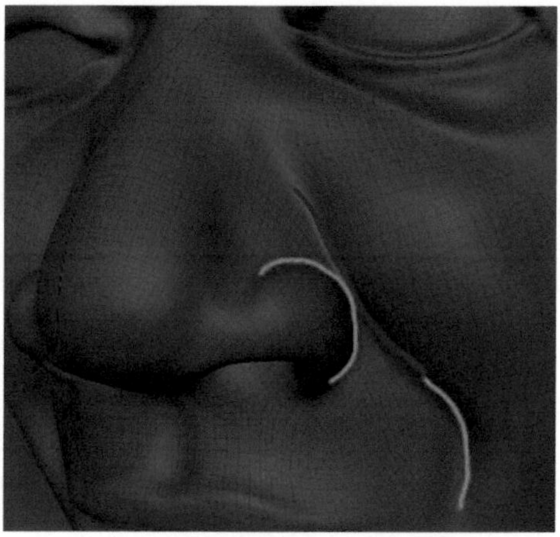

Fig. 3.48 Increased density "forces" the ability to add detail without regard to edge-flow or modeling efficiency. These topics will be addressed in a later stage of the model

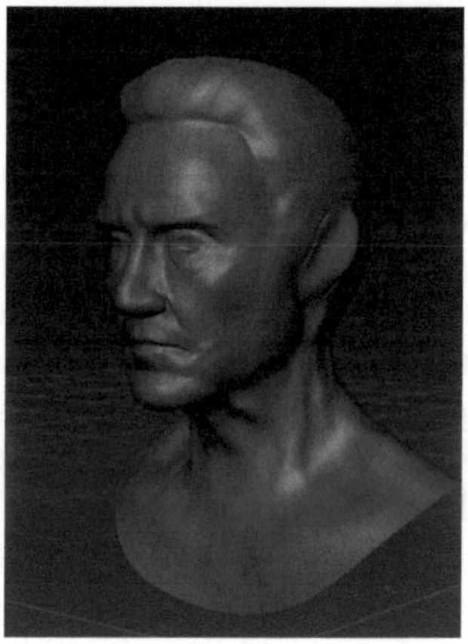

Fig. 3.49 Very high-density sculpture

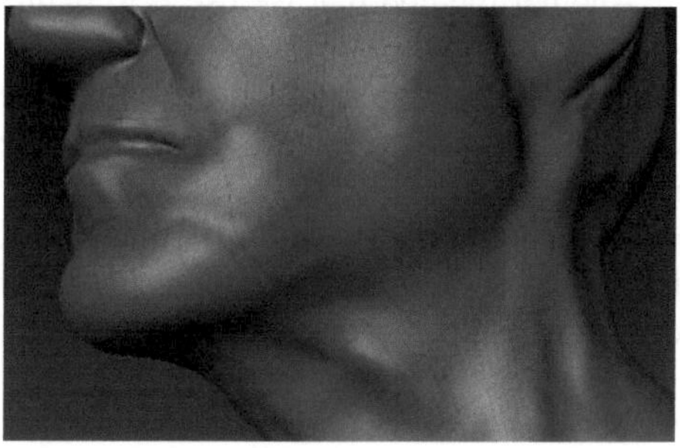

Fig. 3.50 A dense sculpture reveals a high level of detail, and serves as a model for retopology and the extraction of normal and displacement maps

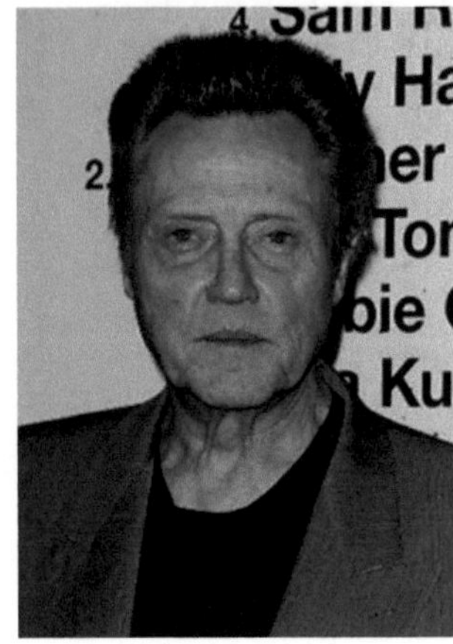

Fig. 3.51 Initial reference image

3.6.1 Topology

As discussed in previous sections, attention to edge flow is critical in defining the various masses and shapes of any object even more so when defining the muscles and structures of the human face. Ridges, depressions, creases, protrusions—all can be very efficiently modeled with relatively few polygons if the edges flow naturally along the contour lines.

We take several approaches in strategizing our edge flow. The first, of course, is to go back to the photographs, back to the drawing board, and sketch in the proper placement of these edges. As an iterative process, we usually recommend that our students do this with a pad of tracing paper or a 2D drawing software like Photoshop which allows for multiple layers and variable transparency.

In Figs. 3.51, 3.52, and 3.53, the student has begun to sketch his polygon layout on top of one of his reference photographs. Notice how the edges form concentric circles around the eyes, following the form of the *Orbicularis Oculi,* and around the mouth, following the form of the *Obicularis Oris*. Each line is carefully placed based on the shape of the underlying anatomy (Fig. 3.54).

Another method that we use is to paint over the face, identifying broader facial shapes based on the muscles underneath (Figs. 3.55, 3.56, 3.57).

In addition to drawing and painting over the original 2D photographs, students can use 3D paint programs to lay out their edge flow directly on their sculpted model (Figs. 3.58, 3.59, 3.60).

By using a combination of line sketches and painting, and comparing those lines with information regarding the underlying structures of the face, the student can then begin the process of retopology. In Fig. 3.53, above, we see different students taking a variety of the recommended approaches in determining the proper layout of the edges. In this way, complex layout "puzzles" can be solved iteratively with a sketchpad before attempting the process using the modeling software.

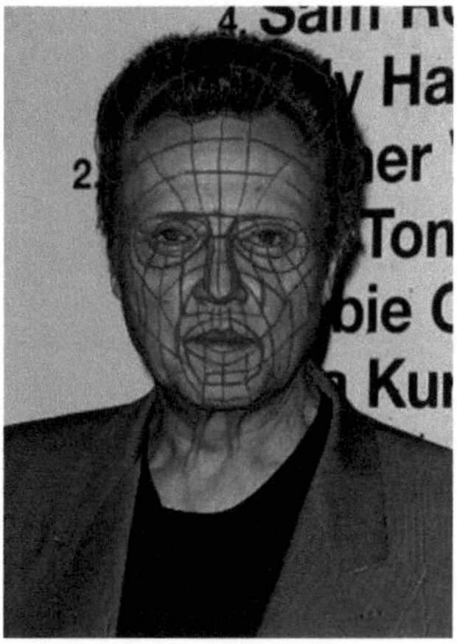

Fig. 3.52 Reference image with edge-flow indicators based on apparent lines and understanding of underlying structure

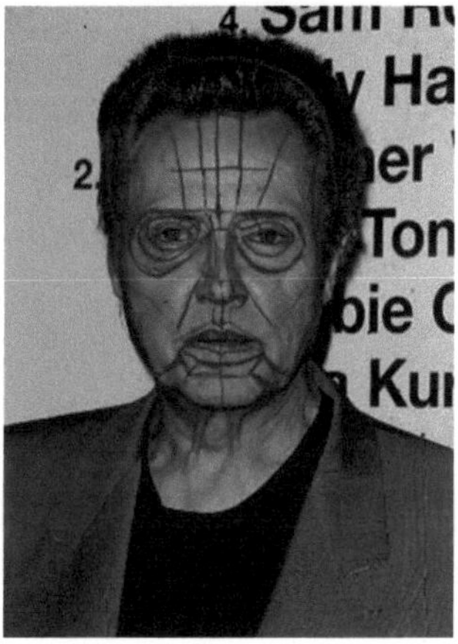

Fig. 3.53 Reference image with edge-flow indicators based on apparent lines and understanding of underlying structure

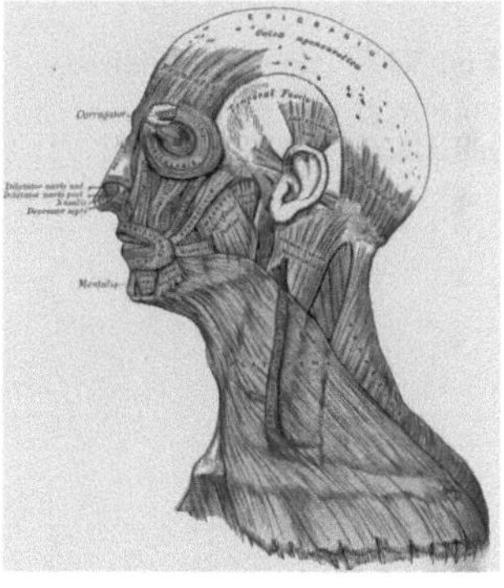

Fig. 3.54 Original anatomical reference

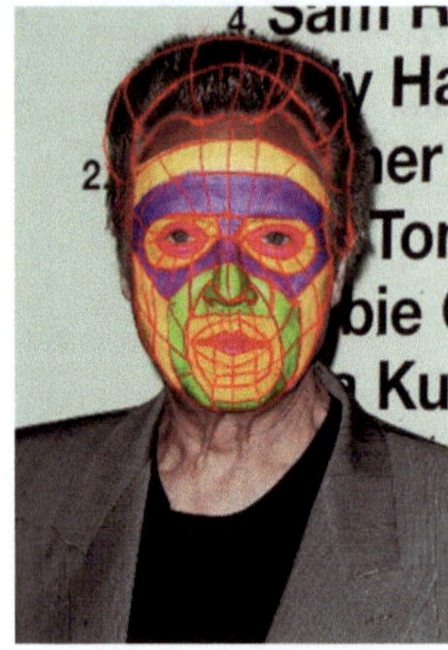

Fig. 3.55 Muscle groups are subdivided into individual polygon edges

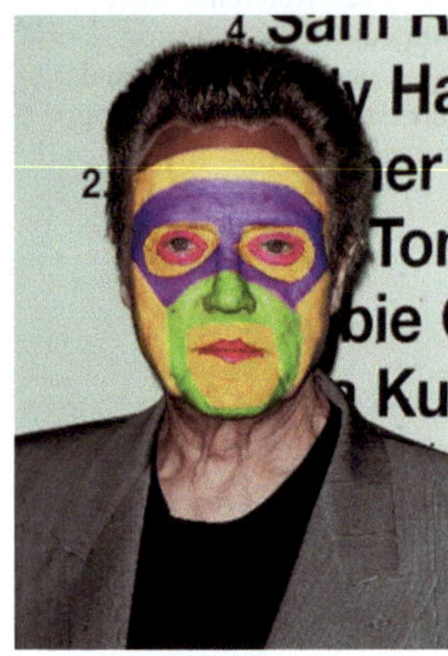

Fig. 3.56 Digital paint is used to define major muscle groups

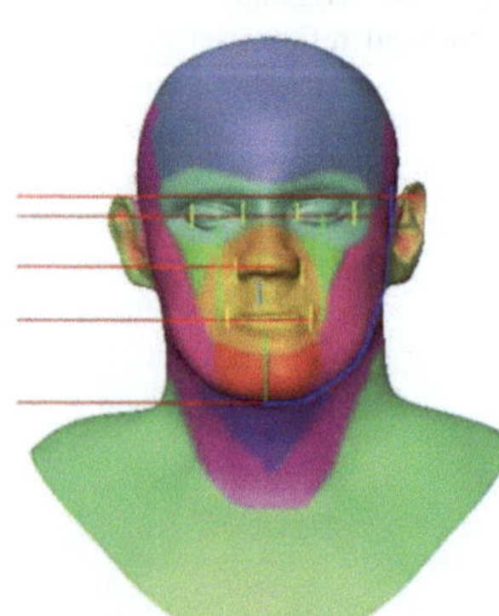

Fig. 3.57 Painting to define muscle groups prior to retopology

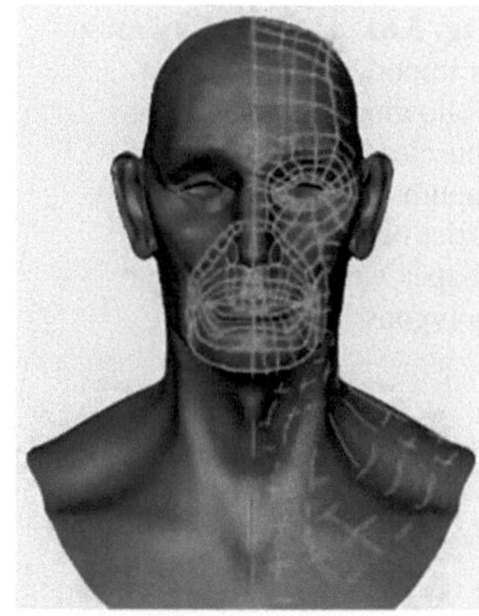

Fig. 3.58 Strategy for retopology of sculpted model

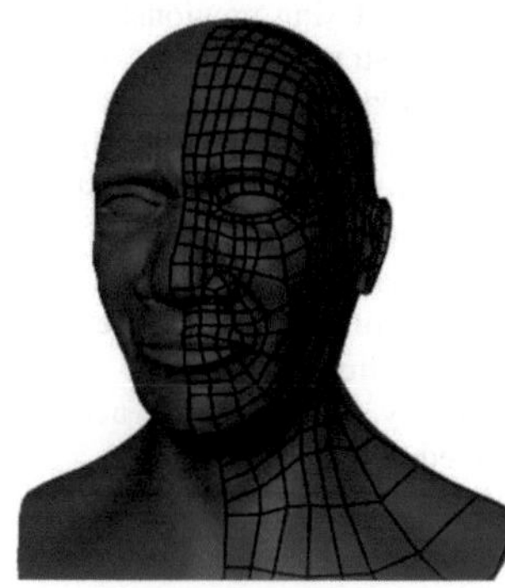

Fig. 3.59 Completed retopology strategy

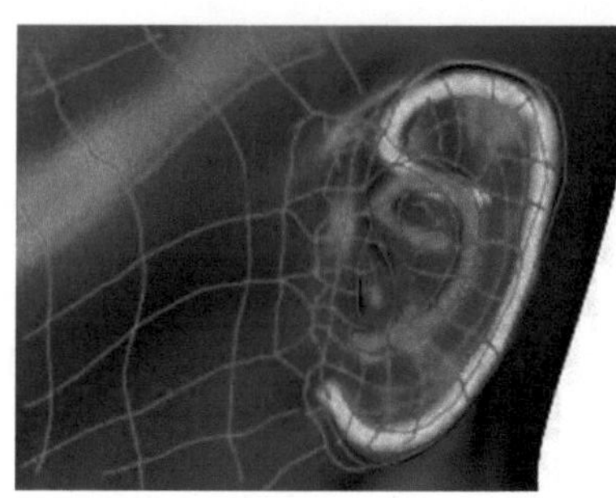

Fig. 3.60 It's often easier to re-do layout details with paint overlays prior to retopology

3.6.2 Retopology

At the time this case-study class was held, we felt that the two strongest standalone retopology packages were "Topogun" and "3D-Coat." Since that time, most high-end modeling software packages have incorporated robust retopology toolsets in their base package.

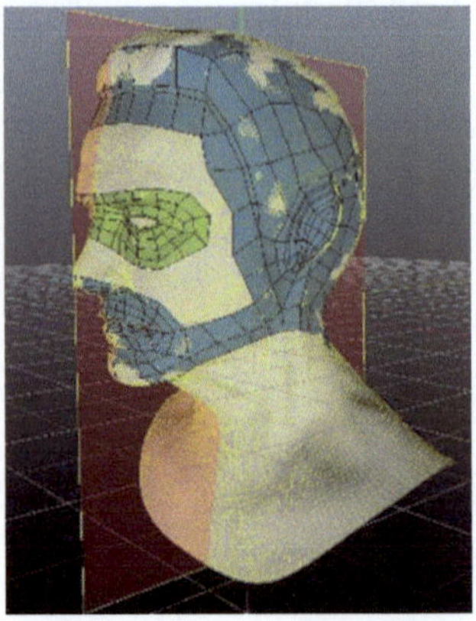

Fig. 3.61 High density mesh is imported into 3D-coat. Following the underlying muscles of the eye, the mouth, and the jawline, the artist begins to trace the shapes with quadrangular polygons

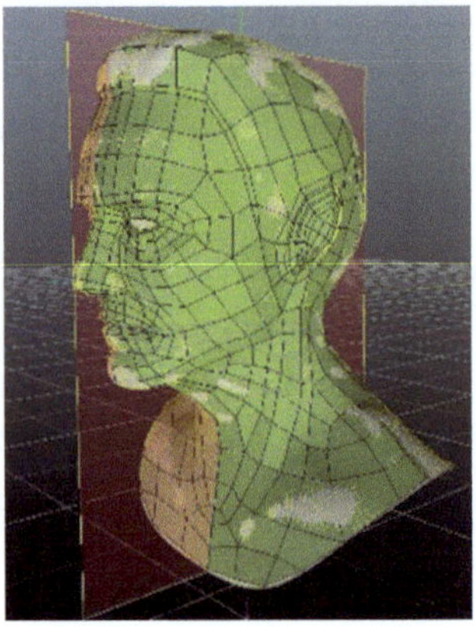

Fig. 3.62 Using previously made illustrations and acquired knowledge of facial anatomy, the artist can finish placing edges where they will define the underlying shapes most efficiently. Note that the 3D-coat software allows the user to define an axis of symmetry, across which each new edge will be mirrored

While the toolsets may have some practical differences, the concept of retopology is consistent. In each case, a high-density base mesh (like the ones we created, or as created using a Photogrammetry process) is imported into the software where it is traced using polygons. Our students have had plenty of practice doing this—the placement of polygon edges follows the same workflow as the draw-overs we have done on paper, in Photoshop, or in 3D paint packages. As we draw with our mouse or stylus, we are drawing with edges and quadrangles. In this manner, we can follow the sculpted forms exactly, while focusing our attention on the technical elegance of good topology and edge flow.

This is the critical point described at the beginning of this chapter; up to now, we have focused on anatomy, on sculpture, on expression—with no regard to any of the technical requirements of efficient modeling. At this point, we can set the aesthetics aside and focus on technology (Figs. 3.61, 3.62, 3.63).

It is important to reiterate the importance of finding the right cutoff, or stopping point, in the initial sculpture phase. In this example, the student may have gone a little too far in adding finer detail to his sculpture before retopologizing; notice in

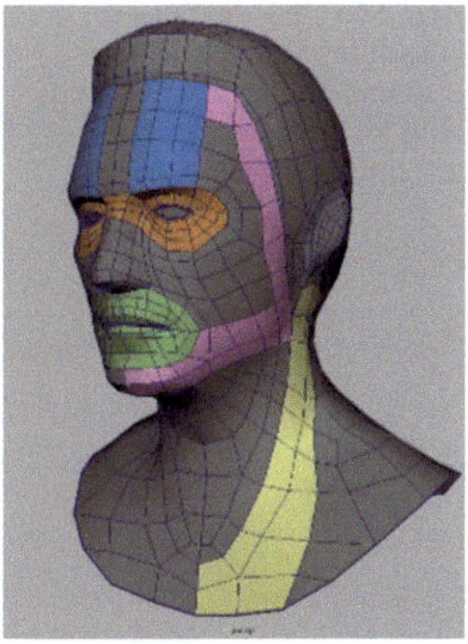

Fig. 3.63 The retopology is completed; the finished model can then be exported for further sculptural detail. Notice how the flow of edges and polygons match the Obicularis Oculi (*orange*), the Obicularis Oris (*green*) the Occipitofrontalis (*blue*), the mandible (*pink*) and the Sternocleidomastoid (*yellow*). The edge flow on these defining facial "landing zones" is critical

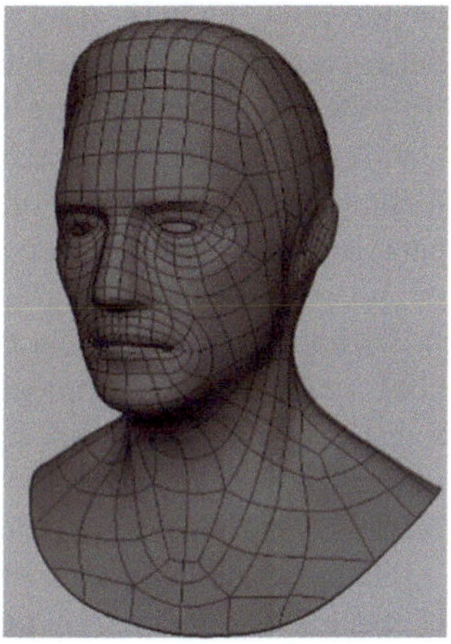

Fig. 3.64 Low resolution, retopologized geometry

the below illustrations that while the main facial structure is intact, much of the finer detail has been lost and will have to be redone. The new, retopologized model can now be reimported into the sculpture package where the additional detail can be readded. With the new, efficient edge flow, this phase of sculpture will be much more easily accomplished (Figs. 3.64, 3.65).

3.7 Digital Sculpture: Case Studies Using Pixologic ZBrush

As the software for organic modeling has become more sophisticated, and the hardware to support this technology has become both more robust and accessible, the result is that the balance of emphasis on technical processes for end users has shifted from being a cumbersome necessity to being nearly transparent. As we saw

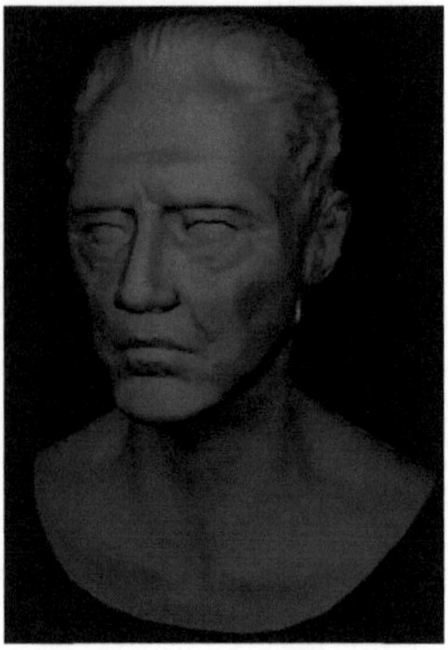

Fig. 3.65 Finished piece; consisiting of the low-resolution geometry rendered with the normal map and displacement map applied

in previous chapters, the "four student" model illustrates this balance/imbalance between technical and aesthetic ability. The ideal student, following that matrix, was the one who had the relatively even distribution of the two skillsets. This sought-after ideal balance has become skewed, with far less emphasis placed on technical ability as a part of the mix. The tools have become so automated that we can focus almost all our energy on design and creativity without the encumbrance of the steep technical learning curve standing in our way.

Perhaps at the forefront of digital sculpture technology is *Pixologic*, a privately held software development company located in California, USA. Pixologic's flagship product, ZBrush, has become a benchmark standard for digital sculpture in the 3D industry and serves as an example of the direction being taken in the processes of both organic and inorganic modeling.

Rather than beginning with meshes that are perfectly topologized and following a sequential subdivision process, the artist can start with a simple sphere and follow a new methodology wherein he need not worry about topology in the initial approach, focusing instead on shape, form, volume, and character. In fact, in the experience of many artists working in the industry, their initial attempts to maintain quality topology from the start frequently need to be redone as the needs of the project or requirements of the art director change. To create the character first and retopologize later is rapidly emerging as a much faster workflow, which eliminates the need to address topology over and over again.

This new methodology reflects some sweeping changes in the industry that are trickling down to education. As much of a shift that the "model—sculpt—retopologize" paradigm has represented in industry and education over the past few years, a case can now be made that traditional modeling, i.e., "box modeling" or any of its offshoots, is becoming more of a preference rather than a necessity. Even the relatively simple step of retopology is frequently considered by some to be a hindrance to spontaneity and the creative process and can now be left to last (and even automated) if necessary or preferred.

From a pedagogical standpoint, we believe it is still important to provide students with a strong foundation in traditional methodologies while keeping a close eye on the industry trends that appear to be long lasting, that will shift the paradigm. Taking the long view of modeling from its beginnings nearly a generation ago, we see this as an important shift in approach that expands, but does not replace, traditional methods.

Of course, in the field of scientific visualization and physics simulation, the underlying mathematics of NURBS curves and surfaces is a necessary part of the design process, but the kind of design that relies on this accuracy is technical in nature. From a purely visual design standpoint, the process of dealing with curves to derive surfaces to attach to create forms is labor-intensive and has become a legacy process, no longer a viable option for production. The same can be said for the additive modeling methods that incorporate a "topologize as you go" workflow. The constraints of the topology are now being replaced by sophisticated, efficient software algorithms that do that work behind the scenes, freeing the artist to stay focused on creativity.

3.7.1 Little Power

In his piece "Little Power," (Fig. 3.66); Mariano Steiner of Pixologic demonstrates the use of a process in ZBrush using a module called "Dynamesh." Starting with a sphere, he is able to quickly grab pieces of the sphere and with using only a small handful of sculpting tool, can pull, push, carve, and otherwise massage this geometry into a very sophisticated shape.

In the first three panels, a piece of this sphere has been stretched downward to form the beginnings of the body structure. A close look at the topology reveals that the edges are initially very distorted and stretched out, but soon become an evenly sized array of fine quadrangular faces appropriate for continuing to add more and more detail to this piece. A feature of Dynamesh is that it creates all quadrangles and triangles which are all the same size, which makes it an excellent tool for fine detail sculpting.

The change in the distribution of these faces in ZBrush is almost completely automatic. At whatever point in the process the artist feels it is necessary, he can select the model and select a menu item which quickly and automatically applies this process to the model.

The first three panels show the topology on the red model, but the artist does not need to see it at all. The process shown on the remaining five panels is one that a traditional sculptor might take with a lump of clay—forming, shaping, and carving—and in some cases, "slapping on" additional bits of material which are quickly and seamlessly incorporated into the model. In many ways, this process is easier than working with clay—the artist does not have to worry about his material drying out, he has an unlimited number of "undo" steps and needs not worry about structural integrity or ensuring that his piece will hold together or not fall. While it

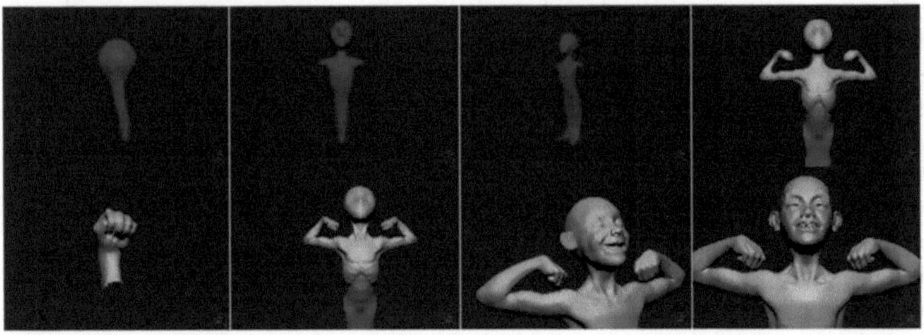

Fig. 3.66 "Little Power"

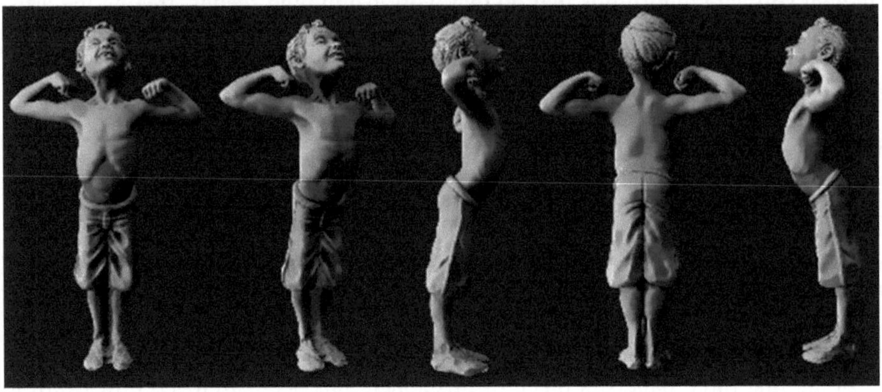

Fig. 3.67 Finished model, "Little Power", digitally sculpted using high-density sculpting techniques

is entirely possible to display the wireframe structure of this model, it is unnecessary to do so at all.

As he works, the artist can quickly add detail to the sculpture by using a density hot key, in "automatic" mode each quadrangle is subdivided once vertically and once horizontally, so each step increases the overall density by a factor of four. There are other ways in which the artist can specify density levels for specific areas of the model as needed—but in either case, the efficiency of the software and the ability of the hardware to effortlessly handle millions of polygons enables the artist to add very fine detail to the model, as in the hands, shoulder structure, and teeth of the example shown.

What is interesting about the finished product, shown here in Fig. 3.67, is that Steiner "built in" specific imperfections to the legs and feet, adding to the sense that this piece was created in clay, not pixels.

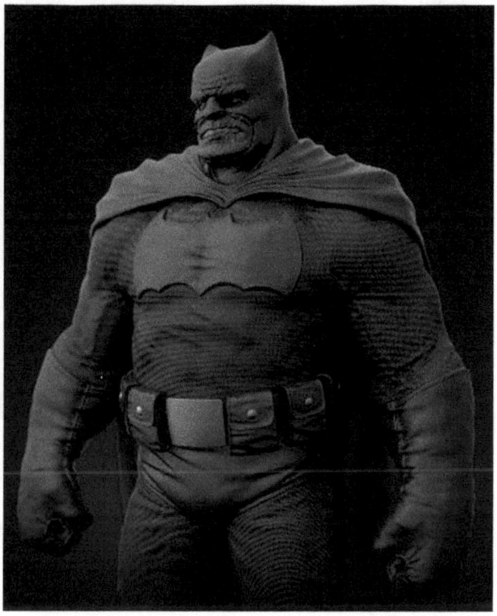

Fig. 3.68 Batman

The intent of Mariano Steiner's piece, "Little Power," was purely sculptural, with little need for a low-polygon-count or efficient edge flow. However, as the next example will show, the additional steps of polygon-reduction and edge-flow optimization for animation are just as seamless.

3.7.2 Batman

Another example of the same initial workflow can be seen in Jason Martin's image of Batman. The original high-resolution sculpture was created following a similar workflow to the "Little Power" in the previous section.

The original sculpture Fig. 3.68 was created using ZBrush Dynamesh and consists of millions of polygons. Using the tools in ZBrush and a subtractive workflow, the artist was able to "decimate" this sculpture, i.e., reduce the polygon count by substantial percentages.

The appropriate number of polygons in the end result is variable; it depends on whether or not the final piece is for use in a game, a movie, or for a 3D print. Further, retopology and reduction may be necessary at this point. Rather than use the decimation workflow, however, which would reduce the polygon count further, a different tool is employed called ZRemesher.

ZRemesher is a tool that will retopologize a model with close attention to edge flow, as discussed in earlier sections of this chapter. This software can do this completely automatically, but if the user has specific areas wherein he wants to dictate specific edge flow to bias the automated retopology process, this

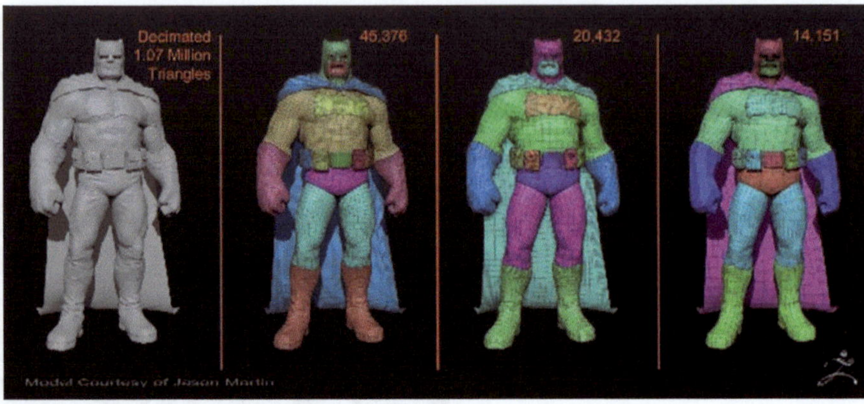

Fig. 3.69 "Decimating" a high resolution model

functionality is also built it. Rather than draw in the quads one by one; however, the user can draw lines similar to those in the pre-modeling draw-overs.

In Fig. 3.69, the subtractive method used on this model is shown. The first image is the decimated sculpture containing a little more than a million polygons; by indicating specific lines on this model to guide the reduction edge flow, the artist is able to reduce this model to several stages of polygon count.

Although detail is lost as the count gets lower, we saw in a previous section how by generating a normal map from the high-resolution geometry, the detail can reappear in the render.

Seeing the results of this workflow, and observing the ease with which these sculptures are created, it is easy to see how this process is "The new wave" in modeling.

Rather than beginning with meshes that are perfectly topologized and following a sequential subdivision process, the artist can start with a simple sphere and follow a new methodology wherein he need not worry about topology in the initial approach, focusing instead on shape, form, volume, and character. In fact, in the experience of many artists working in the industry, the initial attempts to maintain quality topology are frequently required to redo this step as the needs of the project or requirements of the art director change. To create the character first and retopologize later is rapidly emerging as a much faster workflow, which eliminates the need to address topology more than once.

During the process of digital sculpture, the artist will push, pull, inflate, and otherwise stretch the geometry to achieve the expressive forms desired. This naturally results in stretched topology making the addition of details very difficult. Even after multiple subdivisions, this stretched topology will ultimately result in an uneven and unacceptable distribution of detail across the model.

ZBrush's mode, "Dynamesh™" is an automatic retopologizer for the purpose of sculpting. It uses only triangles and quadrangles which are all equal size, which is ideal for digital sculpture.

After stretching the model as described above, the user need only applies the "Dynamesh™" tool as a quick step in the process to quickly and accurately retopologize the stretched areas.

Surface topology, a necessary technical requirement of this process, has traditionally been a natural constraint to the workflow of artists; by eliminating the need to address it so early in the process, and by automating it almost entirely at the end, artists can be considerably freer to explore ideas in effect by "sketching" directly in the software.

Indeed, the need for a strong understanding of the technology has been greatly reduced for the end user; most of the meticulous attention to mathematical surface calculation, surface parameterization, and tangent continuity at shared edges, "edge loops" and edge flow is now taken away from the process by the combination of software, hardware, and the accessibility of both. As we said at the beginning of this chapter, artists are free to be spontaneous and creative without the constraints represented by a steep and cumbersome technical learning curve.

Image Credits

Figure 3.1 "Grays Anatomy"
Figure 3.2 Photo courtesy of www.3d.sk; drawover illustration courtesy of Madeline Rabil
Figure 3.3 Photo courtesy of www.3d.sk
Figure 3.4 Photo courtesy of www.3d.sk; drawover illustration courtesy of Madeline Rabil
Figure 3.5 Photo courtesy of www.3d.sk; drawover illustration courtesy of Madeline Rabil
Figure 3.6 Illustration courtesy of Madeline Rabil
Figures 3.7, 3.8, 3.9, 3.10, 3.11, 3.12, 3.13, 3.14, 3.15, 3.16, 3.17, 3.18, 3.19 Autodesk screenshots reprinted with the permission of Autodesk, Inc.
Figures 3.20, 3.21 Image courtesy of Tony Williams
Figures 3.22, 3.23, 3.24, 3.25, 3.26, 3.27 Autodesk screenshots reprinted with the permission of Autodesk, Inc.
Figure 3.28 Illustrations courtesy of Deanna Giovinazzo
Figure 3.29 Photo courtesy of www.3d.sk; drawover illustration courtesy of Hunter Baule
Figure 3.30 Photo courtesy of www.3d.sk; drawover illustration courtesy of Hunter Baule
Figure 3.31 Image courtesy of Deanna Giovinazzo
Figure 3.32 Karl Langer, "Zur Anatomie und Physiologie der Haut. Über die Spaltbarkeit der Cutis". (1861)
Figure 3.33 Photo courtesy of www.3d.sk
Figure 3.34 Karl Langer, "Zur Anatomie und Physiologie der Haut. Über die Spaltbarkeit der Cutis". (1861)
Figure 3.35 Photo courtesy of www.3d.sk
Figure 3.37 Model courtesy of Eric Kimberly
Figure 3.38 Autodesk screenshots reprinted with the permission of Autodesk, Inc.

Figures 3.42, 3.43 Model courtesy of Eric Kimberly
Figure 3.45, 3.46, 3.47 Autodesk screenshots reprinted with the permission of Autodesk, Inc., sculpturecourtesy of Ben Frazier
Figure 3.49, 3.50 Autodesk screenshots reprinted with the permission of Autodesk, Inc., sculpturecourtesy of Eric Kimberly
Figures 3.51, 3.52, 3.53, 3.54, 3.55, 3.56 Illustrations courtesy of Eric Kimberly
Figure 3.57 Image courtesy of Daniel Pise
Figure 3.58 Image courtesy of Benjamin Frazier
Figure 3.59 Image courtesy of Andrew Turci
Figure 3.60 Image courtesy of Benjamin Frazier
Figures 3.61, 3.62, 3.63, 3.64, 3.65 Image courtesy of Eric Kimberly
Figures 3.66, 3.67, © 2013 Mariano Steiner, image courtesy of Pixologic, Inc.
Figures 3.68, 3.69, © 2013 Jason Martin, image courtesy of Pixologic, Inc.

Chapter 4
Photogrammetry and Design Workflows

Abstract This chapter illustrates a new paradigm in realistic 3D modeling; combining the ability of hardware and software to handle millions of polygons with the ability of digital photography and photogrammetry to produce these high-density models. Various processes are now accessible to smaller studios and universities, and this new paradigm is further easing the technological constraints presented to digital artists.

Keywords Photogrammetry · Autodesk 123D catch · Scan · Scanning · Modeling Toolkit · Maya · Quads · Subdivisions

4.1 Photogrammetry Workflows

Up to this point, our discussions have revolved around user-generated models created by using various traditional and next-generation techniques. We have shown how by following either a more traditional (additive) or next-generation (subtractive) workflow, artists are able to visualize their imagined characters and creatures with relative ease, for film, games, or any number of other applications.

The beginnings of this workflow do not, however, need to start in the computer. Traditional sculptors, working in traditional materials, have always been a part of any visualization pipeline. Better funded operations have long been able to use high-end 3D digital scanning equipment to digitize these physical models for their 3D artists to reproduce in the computer, either using them as modeling reference or going through the (previously) painstaking operation of retopology. As we have seen in the previous chapter, softwares, such as ZBrush ZRemesher, Autodesk Maya, 3D-Coat or Topogun, have substantially streamlined this process.

On the other end of the pipeline, the "input" end, the hardware, the software, and the actual models have also become more accessible to smaller studios and to universities. Following the matrix outlined at the beginning of Chap. 3, the net result is that digital artists are able to fast-track the difficult and highly technical

S. Foster and D. Halbstein, *Integrating 3D Modeling, Photogrammetry and Design*, SpringerBriefs in Computer Science, DOI: 10.1007/978-1-4471-6329-9_4,

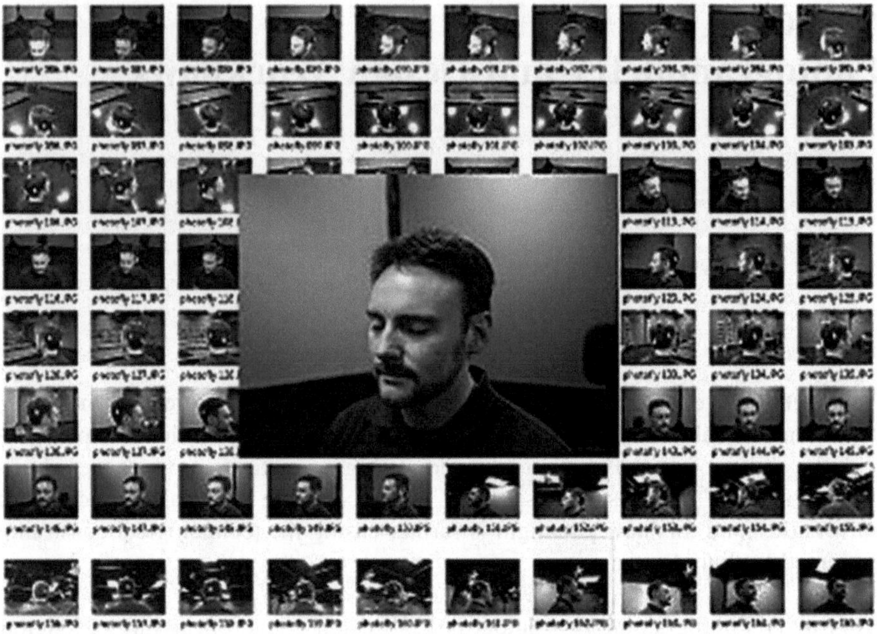

Fig. 4.1 The subject is posed sitting in a chair in a reasonably well lit room

processes involved in model-making, and focus on the creative and imaginative parts to which they may be better suited.

One example of how mainstream this process has become is the development of Autodesk's "123D Catch" software. The software is free and cloud-based; users download a small-footprint image organization program that allows for quick file transfer between their computers and the Autodesk servers.

Using a simple point-and-shoot camera, users photograph a subject from multiple angles, assembling a series of 40–60 images. There are a number of parameters that need to be followed in order to produce a successful file, but the parameters are relatively easy to follow.

In Fig. 4.1, the subject is posed sitting in a chair in a reasonably well lit room. Little attention is given to the lighting, however—no more than standard room light is necessary for this capture. Note, however, that the light is constant, there is detail in the room in the background, and the subject actually has sticky-notes affixed to his head. This photogrammetry algorithm compares points in each shot through triangulation; therefore, it is important to give the software enough information to work with. Foreground and background areas are tracked from one viewing angle to the next—so the subject must be still and the camera moved around him.

This particular session used 80 individual photographs, as shown in the image.

Once these images are compiled into a folder and uploaded, an initial 3D model is returned as shown in Fig. 4.2. The representation of the subject shown in the

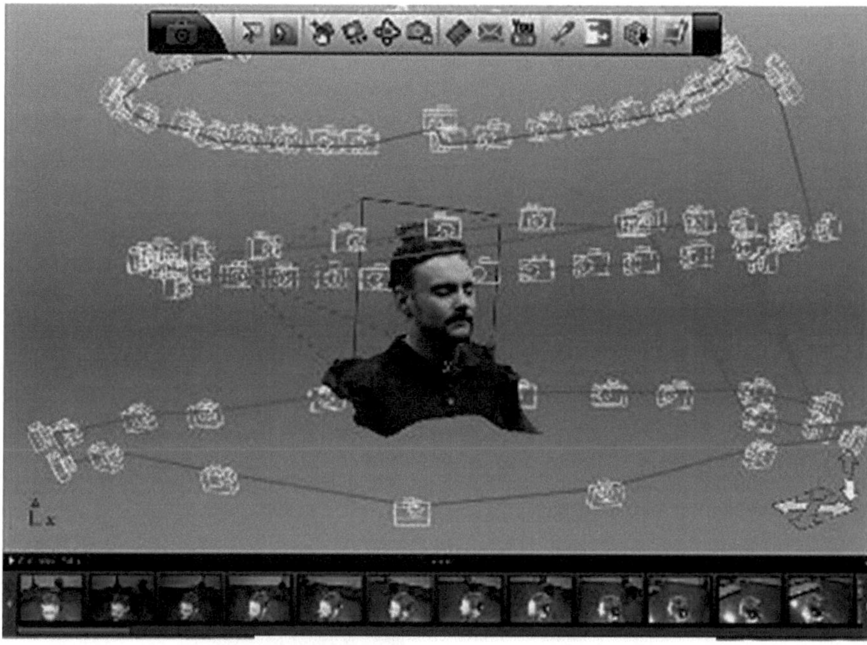

Fig. 4.2 An initial 3D model Autodesk screenshots reprinted with the permission of Autodesk, Inc

center of the image is an actual 3D model; the camera icons can be selected and reference the associated still image at the bottom row.

This is an intermediate phase, showing the results of each calculation. Users can make modifications to the model by deleting unneeded faces and can request that the final calculation be performed at different resolutions.

The texture on the surface of the model, rather than using a standard UV layout, uses a camera-projection method. The individual photographs are broken up and blended, as seen in Fig. 4.3, with each piece being projected onto the model from the angle of the original image. Figure 4.4 shows the final mesh resulting from this process.

This software is very mainstream; its popularity is due to the price (free), the ease of use, and the immediate turnaround of a very good-looking 3D representation of a subject. While the 3D mesh is very accurate and dense, there is a lot of missing information or information that has been "averaged," making this process a little less suitable for a production pipeline than some other methods. Nevertheless, it has reached a very wide user-base, who are using this technology to create exciting interactive 3D visualizations for the Internet, and making use of Autodesk's 3D print capabilities as well. Captured meshes can be returned to Autodesk for printing in various materials including laser-cut cardboard and polymer plastic.

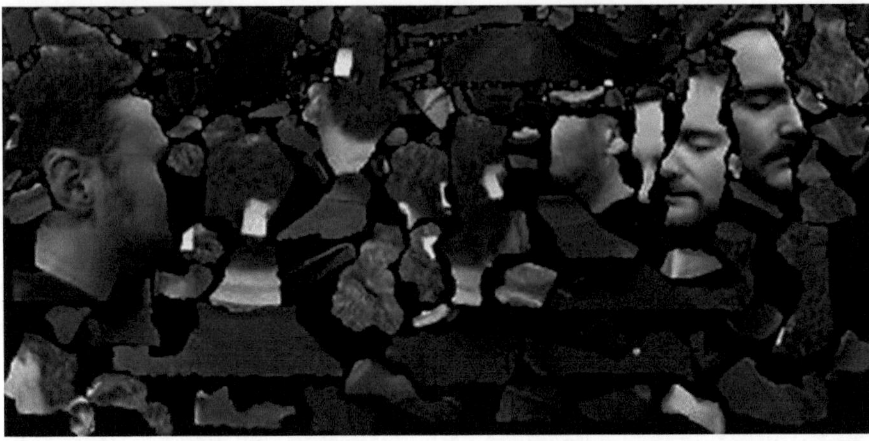

Fig. 4.3 The individual photographs are broken up and blended Autodesk screenshots reprinted with the permission of Autodesk, Inc

Fig. 4.4 The final mesh

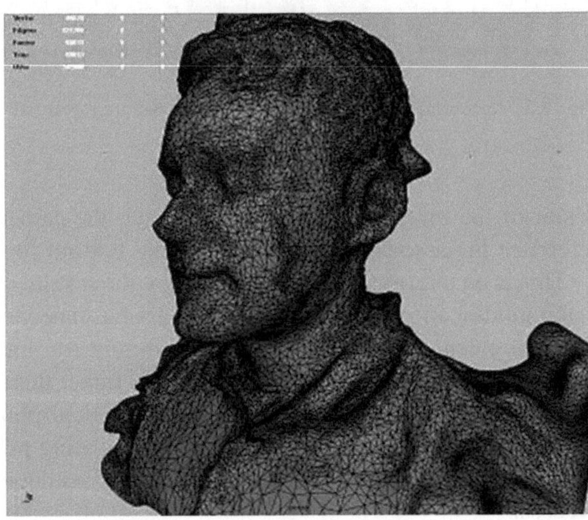

For production-ready scans, it is frequently a better option to use high-resolution scanning equipment, where the finer details can be captured more accurately. While the technology for this kind of scanning remains relatively costly and thus out of reach for smaller studios, independent artists and universities, service bureaus have formed wherein models may be purchased "off the shelf" or custom scans can be contracted. This is a viable option for many and has provided a necessary channel for the "input" side of the production workflow.

One such company is called "Ten 24," located in Sheffield, UK. Figures 4.5 and 4.6 are examples of the kinds of models that can be purchased from them; they

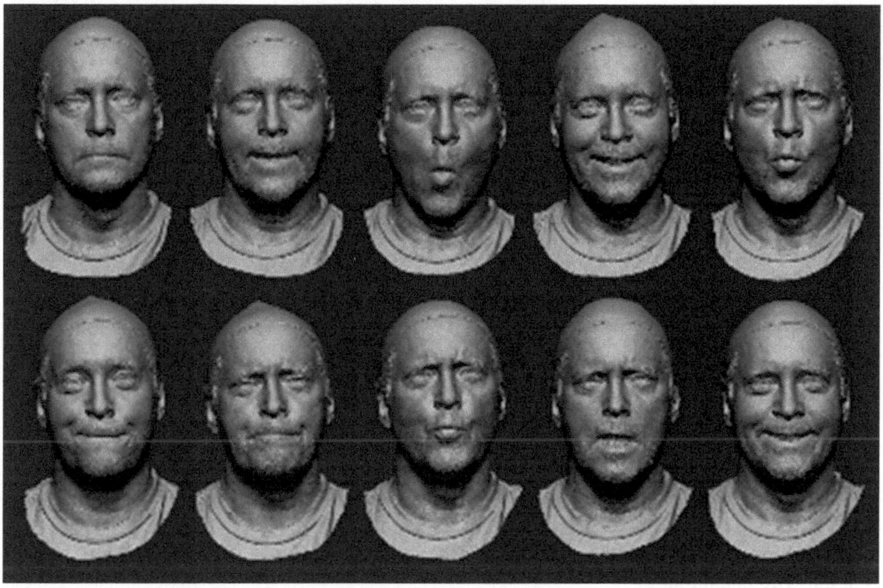

Fig. 4.5 An inventory of scans of individual body parts Images courtesy of Ten24, Inc

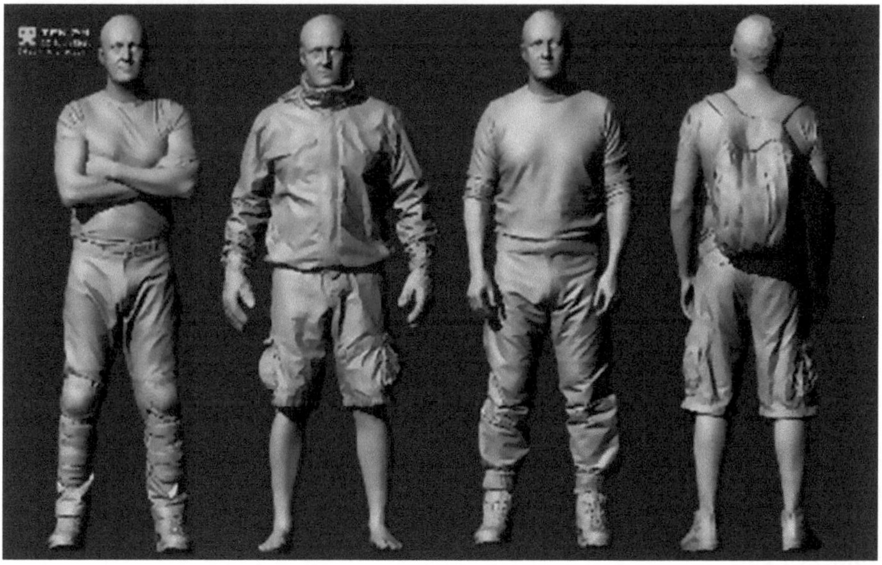

Fig. 4.6 An inventory of full body scans Images courtesy of Ten24, Inc

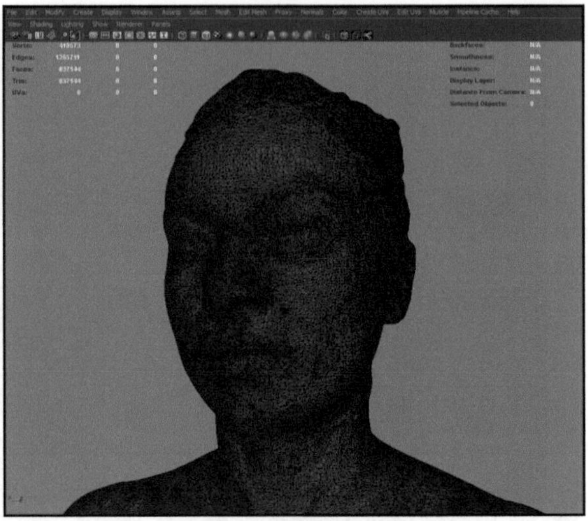

Fig. 4.7 The mesh derived from one of the scans purchased from Ten 24

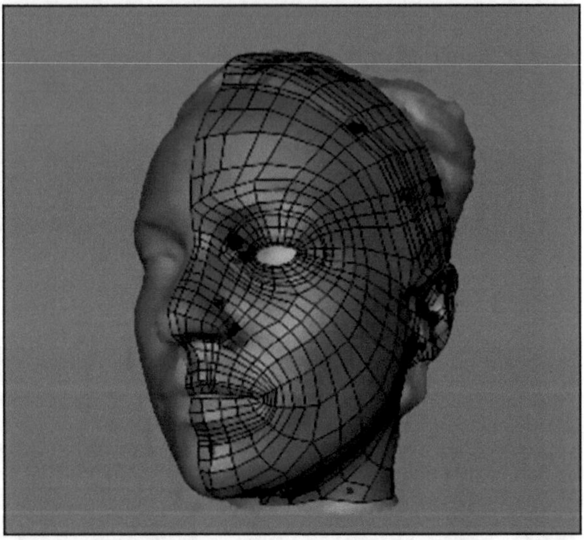

Fig. 4.8 The beginning of the retopology process using Autodesk Maya 2014s new "modeling toolkit"

maintain an inventory of full body scans and scans of individual body parts. Individual and commercial licensing is available for these scans.

Figure 4.7 is an example of the mesh derived from one of the scans purchased from Ten 24.

The mesh is very dense, over a million polygons, but can be utilized in any of the ways described in Chap. 3.

The image in Fig. 4.8 is the beginning of the retopology process using Autodesk Maya 2014s new "modeling toolkit". The underlying scan data can be

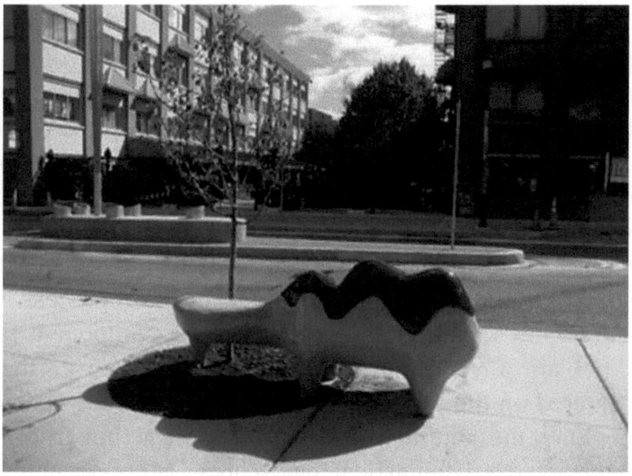

Fig. 4.9 Roberley bell sculpture

traced by the user using a near-standard polygon modeling process. The scanned mesh is made "live" within the software, and the user can simply place points on the surface from which quadrangular geometry can then be derived.

It is important to have a clear idea of the intended edge-flow layout before beginning this process and to adhere to the plan. For this reason, it is equally important (as described in Chap. 3) to strategize this edge flow using a 2D illustration program like Adobe Photoshop, or even prints of the image with colored pencils. Once the user has acceptable drawn-over lines, he can then begin the process of applying the quads.

As points are added and quads are subdivided, the new subdivisions automatically update to adhere to the underlying surface.

At points throughout this process, it may be necessary or desirable to exit the modeling toolkit and apply more traditional polygon modeling techniques—to fill holes, or to massage or soften edges, or to split individual polygons. Once this has been done, the work in progress can be selected and "shrink-wrapped" to fit the scanned model. In fact, primitive objects such as cylinders can be located around large areas on the target surface and "shrink-wrapped" to fit. This workflow is well suited for beginning to model arms and legs if not too much attention to edge flow is required.

Once the base mesh has been retopologized, the scan data can be set aside and further sculpting can be done, normal and displacement maps can be extracted, and the processes outlined in Chap. 3 can be followed (Fig. 4.9).

4.2 Case Study: Roberley Bell

Sculpture at the intersection of the hand the machine-made.

For commercial interests, a high level of control is necessary with 3D photogrammetry in order to achieve a predetermined output. With artist Roberley Bell, she has an artistic process which explores and participates with some of the aesthetic "contributions" that the technical process delivers as part of the final product.

At what point does strangeness in scanning become an interesting part of a creative, finished product? The Japanese "Wabi-Sabi" aesthetic is a philosophy which centers on the acceptance of imperfections [1]. Often this aesthetic actually values what others would see as imperfections. In a recent artist statement (below), she discusses how her technical and creative process has led her to the point where she is now trying photogrammetry as a component to her designs.

"I am a visual artist an object maker. Over the past decade my work has centered on the production of sculpture and site specific public projects. My work explores the natural world both in abstracting from, and in borrowing, to reveal hyper-realized fantastical landscapes. My current series *Other Landscapes* fast-forwards the origins of organic abstraction into the twenty-first century. My forms take their cue from Blob Design. Like the current trend in design where buildings and form adopt an organic structure that is made possible only though computer-aided design, my *Other landscape* sculptures, reveal themselves as natural forms, though they are, in fact paradoxically based on nothing that exists in nature.

I have recently begun incorporating new technologies into my hand making process of form. Utilizing a CAD (computer aided drawing) program rendering blob forms via the computer and sending them to a CNC router (computer controlled shaping machine.), which then digitally shapes the foam substrates of my sculptures. Beginning with watercolor sketches and small clay models I then, with the aid of a fabrication shop, draw the forms in a CAD program they are outputted in foam via a CNC process at this stage I return to the hand making process working the surfaces to articulate the forms, the nuance of the sculptures surface. This evolution of my process engages innovation and extends the creative studio practice beyond the tradition of hand working."

4.3 Case Study: Sucker Punch Studio's

4.3.1 3D Scanning of Faces and Clothing for "Infamous: Second Son" Game

https://www.gameinformer.com/games/infamous_second_son/b/playstation4/archive/2013/05/24/how-sucker-punch-photocopied-actors-for-infamous-second-son.aspx

The art style of the game Infamous combines superhero powers with a very realistic aesthetic. In a trailer released for 2013 E3 (http://www.youtube.com/watch?v=MINfjvFnzc8), they revealed clips from their upcoming game. In addition to the beautiful graphics, one of the most striking visual aspects was the smooth and natural character facial movement. Sucker punch studios not only used

3D scanning for pedestrians and clothing, but the biggest application was for facial animation. According to Max Chapman of Sucker Punch[1]:

"The real reason we adopted scanning...was for our facial system. In terms of assets the real benefit is quick and efficient data to either work from, or directly adapt to increase realism, fairly straightforward, but not always simple. In the greater scheme of things it's more cost effective" he also went on to say that "it's time consuming and quite complicated."

Sucker Punch's facial scanning, referred to by the acronym FACS, (facial action coding system). Allows Sucker Punch to "...scan a host of facial features and use the data to drive complex rigging systems to add complex micro movements in our characters for greater realism. The end result is that Sucker Punch is able to visualize and "basically able to adapt skin, bone, muscle, and fat interactions on a human face. We would have never likely bought this equipment if it wasn't for that."

From your perspective as someone at a top game studio, where do you feel the technology is going?

"As far as scanning it's the hybridization of technologies in the never ending quest for greater realism. Real-time scanning capture solutions replacing or augmenting rather simulation is one of those areas. Higher resolution at faster and faster camera speeds is the big benefit. Basically, how can you get a massive amount of accurate data in real time and use that in conjunction with paradigms that generate usable topology. On top of that there is the issue of capturing accurate texture and light data for robust generation of BRDFs that could also pipe back into real-time solutions. This is a long path of course, but it gets better and better by the year."

How do these tools free and affect creativity, whether it be games or film?

"...There is simply too much of a risk versus profit ratio to overcome that is severely hampering the market as a whole, costs need to get more reasonable or the industry will become unsustainable, meaning innovation/creative risk. At the moment it is a thirst for realism, e.g., war games, GTA, sports, hopefully they technology will help us move beyond that. Scanning is only one part of the equation of course, but all in all technology almost always holds a major part of the key to these issues."

Reference

1. Koren, L.: Wabi-Sabi: for Artists, Designers, Poets & Philosophers. Imperfect Publishing, California (2008)

[1] Interview with Max Chapman, Lead Character Artist, Sucker Punch Studios (October 18th 2013). © Sony Computer Entertainment America LLC 2013.

Chapter 5
3D Design and Photogrammetry

Abstract This chapter describes a framework for the design process with a specific focus on three-dimensional designs. It gives an overview of the elements of design: color, line, shape, texture, space, and form, and then the principles of design: unity, emphasis, placement, size and proportion, balance, symmetry and asymmetry, repetition, proximity, and dynamics. It also investigates convergent design paradigms from cinematography combined with 3D designs, which results in 2D images.

Keywords 3D design · Elements of design · Color · Line · Shape · Texture · Pattern · Space Principles of design · Unity · Gestalt, emphasis · Balance · Symmetry · Asymmetry · Repetition · Proximity · Dynamics · Cinematography · Lens · Camera · Film

Designs are a mix of creative problem solving by applying various principles and tools. There are various *elements* and *principles* of design that can be applied while integrating new visual tools and technical workflows. In this section, we are discussing some of the ways design tools are integrating digital 3D forms to the design process. We will approach the combinations by reviewing traditional design rules by first defining the elements (or components of each image) and principles of design while spending specific attention to ways 3D design can be applied to these works. How elements are used will be more formally discussed in Sect. 5.2. At the end of this chapter, we will expand upon the new workflows and opportunities that are starting to become realized by using the new photogrammetry and retopologizing workflows.

5.1 Elements of Design

The elements of design are visual components that any image can be separated into. Unlike the principles of design, the *elements of design* are almost uniformly agreed upon set of attributes. One major distinction that can be made between

S. Foster and D. Halbstein, *Integrating 3D Modeling, Photogrammetry and Design*, SpringerBriefs in Computer Science, DOI: 10.1007/978-1-4471-6329-9_5,

elements of 2D and 3D design is the addition of (3D) form. For this text, we will discuss: **color**, **line**, **shape**, **texture**, **space**, and **form**. All of these elements of design can *combine* to create imagery in nearly infinite combinations. As opposed to "real life," a design often uses elements in a way which abstracts reality, using either reductive or exaggerated elements. We will discuss each element, defining it, with a specific focus for interpretation and discussion of some of the various common tools in most graphics programs. These elements will be discussed within the framework of how they can fit as component elements in 3D graphics, and photogrammetry-based 3D workflows, as well as rules of thumb for achieving stronger graphical output.

5.1.1 Color

Colors are a breakdown of (visible) light within a small part of electromagnetic (EM) spectrum that humans can perceive. The science and use of color is an extensive topic. This book will give an overview, with a focus on resources for developing color palates for use with the 3D design process.

There are two main ways of defining color space: **additive** or **subtractive** color. Physical media such as paint are subtractive. Light at various wavelengths of color is absorbed by paint and gets reflected back to the eye. Mixing all of the colors subtracts (absorbs) all the color reflected to your eye until you achieve black. For this book, we will be discussing color from the perspective of digital media, which is **additive**. In additive color space, the main (or primary) colors are red, green, and blue. When these primary colors are mixed together, the result is white. In the figure, three colored lights overlap (add) to create "white." The primary colors red, green, and blue are mixed (as shown in Fig. 5.1) with each other to form the secondary colors yellow, magenta, and cyan (Fig. 5.2).

Often combined with *tints* (whites) and *shades* (blacks), the application and choice of color within design has a strong impact on the finished image. Specific combinations of colors, or color palates, are a foundation for many designs.

Color in design is very subjective and will vary by cultural, temporal, and national perspectives (see Sect. 5.1.1.2 below). As the globe becomes more connected, the greater contact with international populations highlights this subjectivity by the exposure to multicultural design as well as discussion of various meanings given to color (see discussion of *Isomorphic Correspondence* in Sect. 5.2 later in this chapter)

Each color can be mixed with various tones (tints and shades) (Fig. 5.3). How the color is affected by the tonal changes depends on the amount mixed with the color as well as the tool and display format used to present the final image.

Levels: This has become a standard tool in many programs. It allows you to specify a point on within the white, blacks, and mid-tones of your image and selectively stretch or contract that value. They are set with the three triangles below the graph. To achieve the widest range of tone for each image, one goal is to

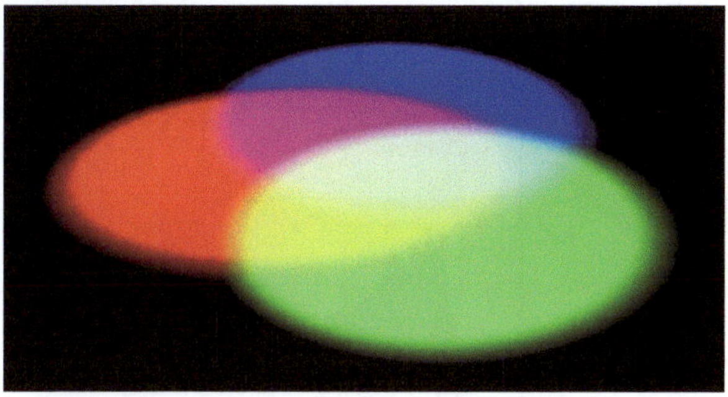

Fig. 5.1 Additive color made with 3 lights in the program Maya

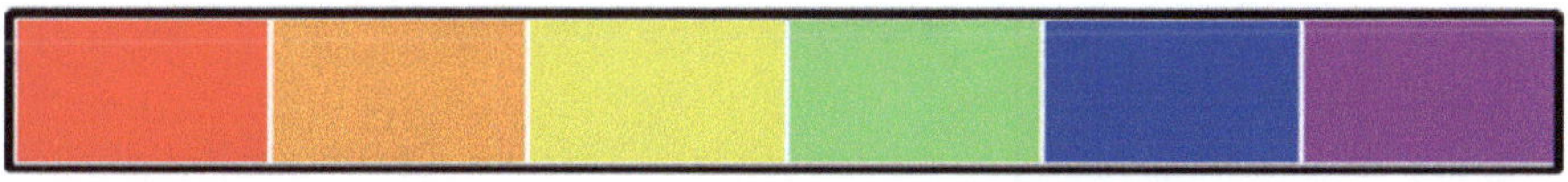

Fig. 5.2 Color strip shows primary and secondary colors

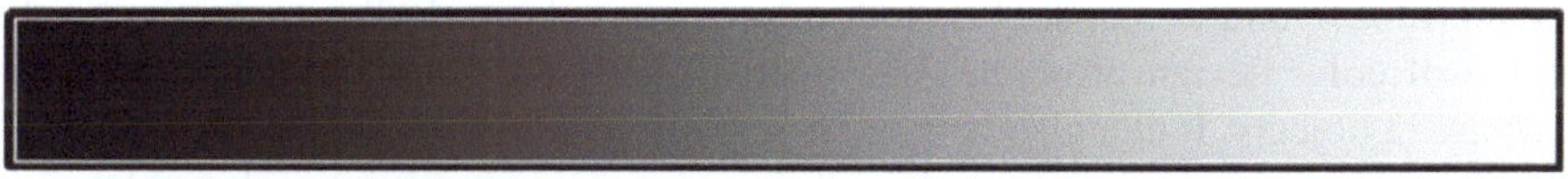

Fig. 5.3 Tone ranges (*tints and shades*) within a color

get the full range of lights and darks a computer can display. Some programs can remap the tones (Fig. 5.4), by adjusting the triangles a user can chose to redefine a greater or narrower range of tones. The dark tones are on the left and the lightest on the right.

5.1.1.1 Relationship to 3D Form

For 3D objects, the color and form are realized by a combination of the object color combined with any color from the light source illuminating that object. In the real world and with some advanced computer-rendering techniques, the color of an object further influences its surroundings by bouncing additional color off of the object as well as potentially transforming light passing through that object (assuming object translucency/transparency).

The perception of depth in 3D forms can be influenced by the use of light and color. How light travels across the surface of an object describes the objects' three dimensionality. How light and color interact can subtly affect our perception of depth. This is due to a biological impact of how the human brain interprets colors and the perception of depth. This can be used stylistically within the 3D design process.

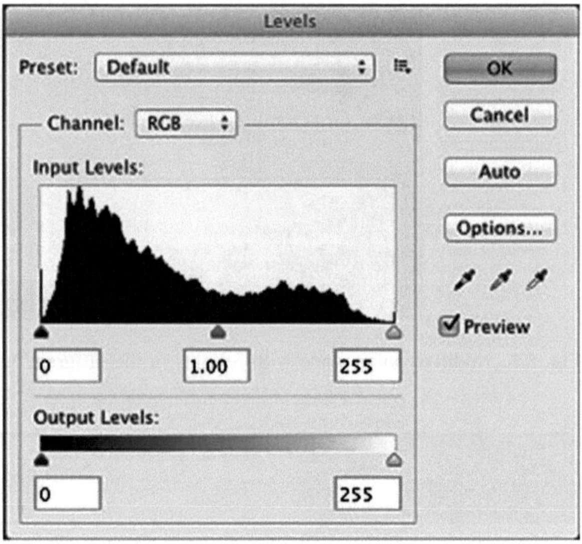

Fig. 5.4 Levels. Adobe product screenshot reprinted with permission from Adobe Systems incorporated

Warm (red–orange–yellow) colors are perceived as "pushing forward," and cool colors (blues) tend to recede. When developing 3D object designs, selective application of color design principles can intensify or reduce the perception of depth within 3D objects. It can also help to increase perception between foreground and background elements. In 2D design, this is done by modifying color palates of foreground and background elements. In 3D design, this can be done in a similar fashion to 2D design, but with additional control of having color within 3D lights affects foreground and background. One example demonstrating this is to set up a scene with a character standing outside. Lights are placed in the foreground which has a slight warmly colored "key" light (yellow or orange hue) directly on the character while having the background lit with ambient color from the sky, with a slight blue hue. The resulting image has the character with slightly more red tones, and this helps give the character greater visibility and visual prominence in the scene. A reverse of this process, casting cool, slightly bluish lights on the main character in the foreground, while having warm colors in the background, can also be applied adding a feeling of reduction or withdrawal to a character.

5.1.1.2 Color Palates

In design, there is often a reduced set of colors used when compared to the broad spectrum available in real life. A color palate reduces the color choices for a composition, to the main colors used (plus black and white). The color wheel (see Fig. 5.5) can be divided in **primary** colors, which are colors which cannot be created by mixing any other colors (red, green, and blue), each primary color, however, can

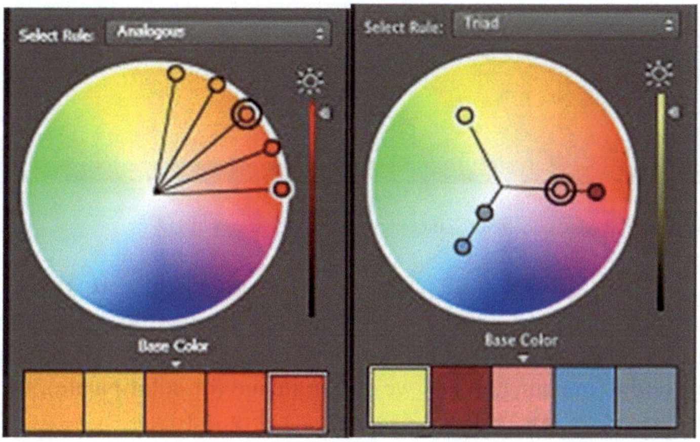

Fig. 5.5 Color palates (Kuler). Adobe product screenshot reprinted with permission from Adobe Systems incorporated

be mixed with another primary color to create the secondary colors (yellow, cyan, and magenta), and also each secondary can mix to create the tertiary colors.

There are various starting points for color palates. Colors across from each other on the color wheel are called "complementary." In the color palate, this color combination produces the strongest visual contrast. Analogous color palates provide a color palate of color on a similar side of the color wheel, potentially helping reinforce visual unity (see Sect. 5.2). Triadic colors are evenly spaced around the color wheel (Fig. 5.5 analogous and triadic).

Joseph Albers was an accomplished artist across several disciplines (typography, photography, and design), and he is best known for his series "Homage to the Square," which investigated color interaction between squares of different sizes and colors. In his book *Interaction of Color* [1] (in 2013, transformed into an interactive iPad application), he described many of the visual and physiological impacts that variation in color and palates has. Many of his writings apply to elements of technical process and design components. The interaction of colors in a design has multiple effects on several levels: biological, cultural/stylistic, and individual significance.

5.1.1.3 Culture, Stylistic, and Individual

The emotional and stylistic perceptions of colors are heavily influenced by the culture a person lives in and also from an individual and temporal perspective. For example, Irish consider the color green to be lucky, while in France, it has an opposite perception. Green also has a strong connection to the religion of Islam. Almost all colors have multiple meanings on cultural, individual, and temporal levels. Often specific color palates will also be popular for some period of time.

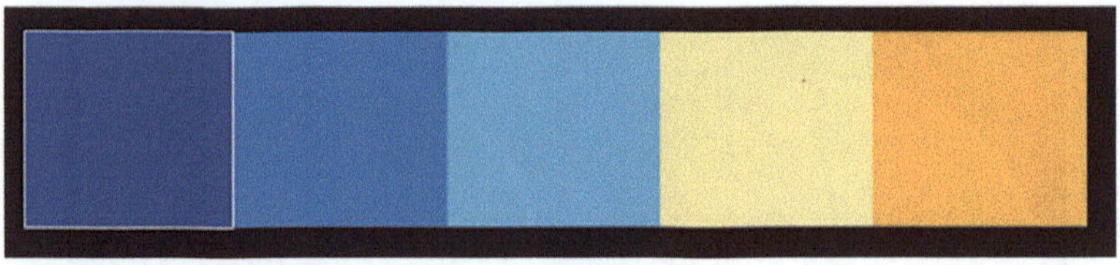

Fig. 5.6 Color palate derived from Vincent van Gogh's starry night

For example, some neon color palates were popular in the 1980s in America, whereas various shades of yellow, green, and brown were popular in the 1970s. Known in Gestalt psychology as Isomorphic Correspondence, this principle states that we respond to meaning. When we see an image (or color palate), we interpret its meaning based on our personal experiences (Fig. 5.6).

5.1.1.4 Color Tools

There are many programs and apps available for developing color palates. One program integrated into multiple Adobe programs is called Kuler. It is available both online at kuler.adobe.com and being integrated as a program extension into many Adobe products (Photoshop, illustrator, etc.). It has specific tools for creating and saving palates, as well as helping to create starting points using color palate rules—for example, analogous, triadic, and complementary. For example, Vincent van Gogh's painting Starry Night was used as a reference to drive a color palette (Fig. 5.6).

5.1.2 Line

Line use within design often stems from analog media such as a pencil or other analog drawing implements. Due to the shift from analog to digital art, there have been attempts to transform approximations of analog tools into digital forms. In physical media, a line is a mark made by a moving point. In analog media, the variation in **direction** and **weight** produces an almost unique look for each artist (Fig. 5.7).

Lines also exist in many ways in nature, for example, lines on leaves, tigers, and on birds. They can serve for opposite purposes also, to make objects stand out or to help camouflage an object with its surroundings.

Our eyes are particularly geared toward understanding lines and using them as part of object detection. Within high-contrast images, objects are often identified by their silhouette. It has been long used within design to use lines to guide the eye to different parts of the image [2]. This has also been established scientifically (Motter, Brad C, and Belky).

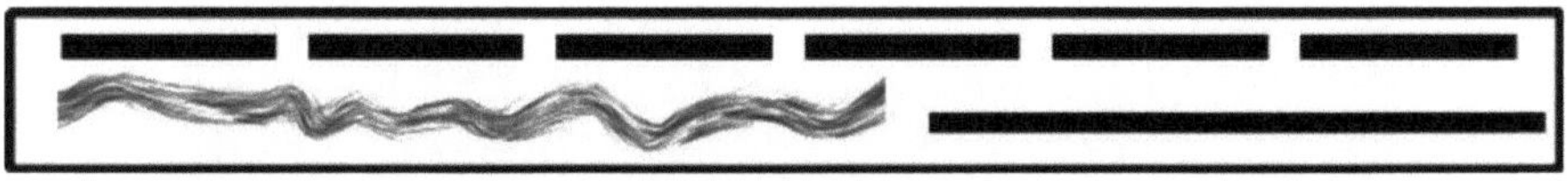

Fig. 5.7 Line

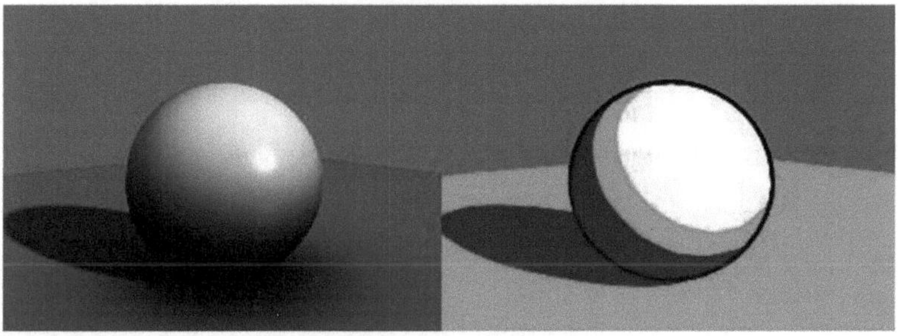

Fig. 5.8 Lines and form

The emotional effect that lines can generate within designs is significant. Horizontal and vertical lines usually communicate stability and solid forms. How the lines are used within imagery further affects the emotional impact. We will discuss this further in the section below on composition.

Lines can be combined with colors and other lines to create textures and patterns (see the Sect. 5.1.4 below for further discussion).

5.1.2.1 Relationship to 3D Form

Many 3D programs often feature the ability to create "toon lines". These lines are created around each object on a 2D plane parallel to the camera (see Fig. 5.8). Computer programs generating images of 3D objects have the ability to extract information about the 3D object, which describes whether the faces of that object are toward or away from the camera. Within the program Maya, the "Sampler Info Node" allows the user to extract the facing ratio. The facing ratio generates a value that is returned from 0 to 1 as to what percentage the plane of geometry is facing the camera. Using the facing ratio to apply color or luminance information to a 3D object which consistently faces the camera is how a custom cartoon silhouette line is created.

When used in combination with a "toon shader," which simplifies lighting, this helps to create the perception that the 3D objects are drawn. Standard black or white toon lines help cartoon objects to stand out from their background while creating a visual style.

A custom toon line can give subtle or strong additional design control to the artist. Lines can be used in 3D design to accentuate or deemphasize the 3D shape

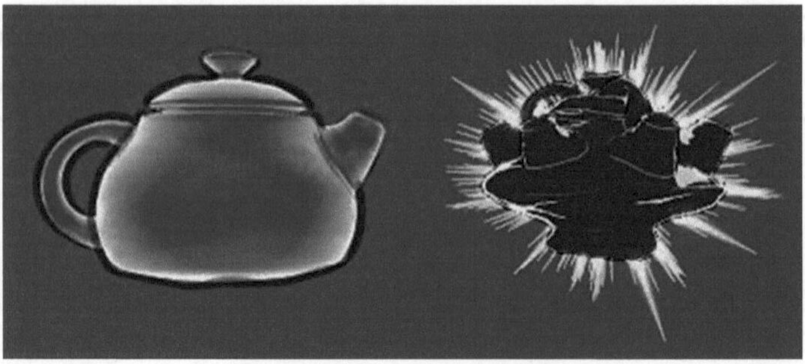

Fig. 5.9 Contrasting and accentuating expressive line and 3D form

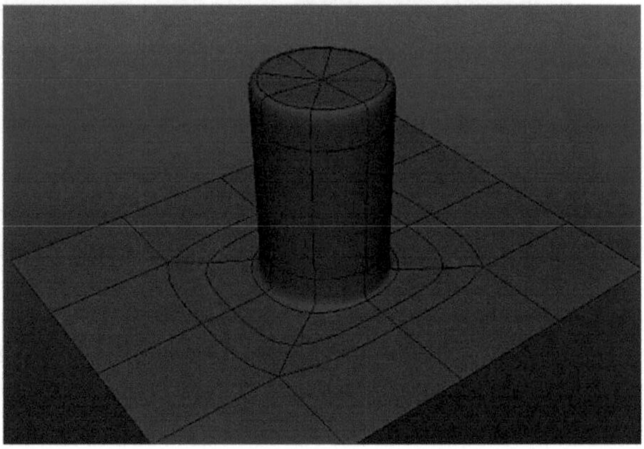

Fig. 5.10 Line flow, underlying structure

of the form. Given that a separate line pass in toon shading can be generated, a line could be used to accentuate curvilinear or rectilinear forms. It could also be given additional "emotional" (examples from Maya Visor, Fig. 5.9) design attributes. Finally, it could also emphasize the medium being used, like the (below left), which gives a water-colored look.

Line direction within the composition can greatly affect the emotional feeling of the image. If the lines in an image are strongly horizontal or vertical, often the image feels static, while diagonal lines often create greater feelings of movement. Strategies for using digitally created lines will be discussed more in the section on composition.

In 3D design, the underlying structure of an object is made-up of connected polygons. The direction and shape of the polygon describe the underlying form of complex objects. In the simple as shown in Fig. 5.10, the lines describe how the cylinder intersects the plane below.

In a more complex example, within humans, there are lines which describe how the skin will fold and deform by default, called Langer's lines. In addition to the skin having default fold patterns, it also follows underlying musculature, bone as well as how the body needs to move. Therefore, the underlying line flow (also known as edge

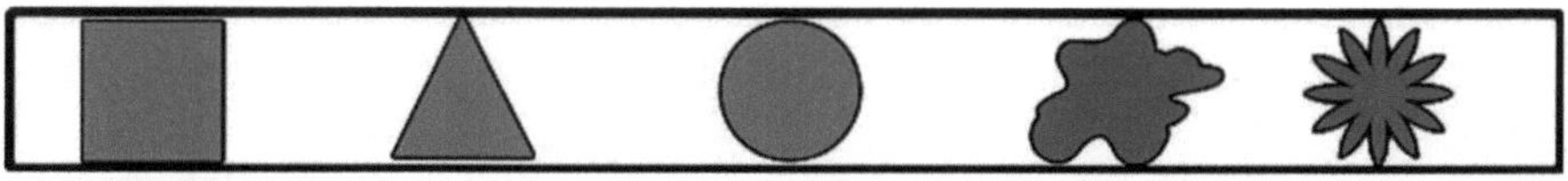

Fig. 5.11 Shape

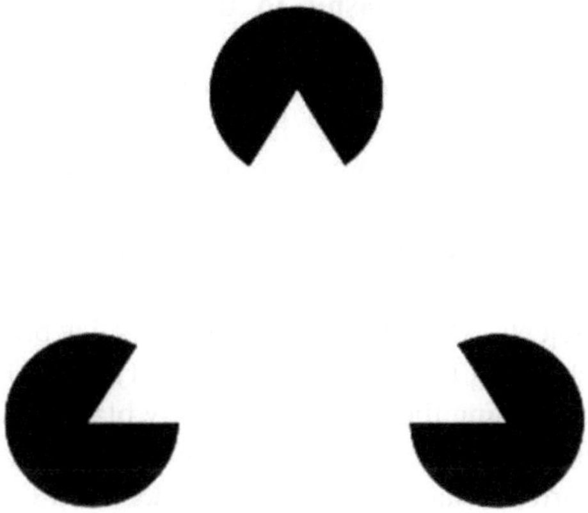

Fig. 5.12 Gestalt reification: implied *triangle* shape

flow) structure of a 3D object needs to reflect how the object will deform. This is one of the reasons for retopologizing objects as discussed further in Chap. 4.

5.1.3 Shape

Shape is a description of a two-dimensional area. Shapes can be created and perceived in multiple ways. Often they are described by a closed group of lines (as the black outline identifies common shapes in Fig. 4.11); however, shapes can also be defined by a commonly filled area. As in Fig. 4.11, each shape is filled with gray color. Shapes can be typically categorized into two categories: geometric shapes or organic shapes.

Geometric shapes are easily described mathematically. They are regular or precise and are regularly found in man-made (3D forms and) objects. In Fig. 5.11, the first three shapes, square, triangle, and circle, are geometric shapes.

Organic shapes are irregular and difficult to describe with simple math and are often found in nature.

Puddles, leaves, flowers, clouds, roots, rocks, trees, etc., and the final two shapes in Fig. 5.11 (right) are all examples of organic shapes.

Shapes can also be implied by the use of negative space (see Fig. 5.12 Gestalt reification). Implicit sense of line and shapes can be perceived within a negative form. The Gestalt principle of **reification** is the term used for the brains ability to construct or generate shapes and forms inside of negative space.

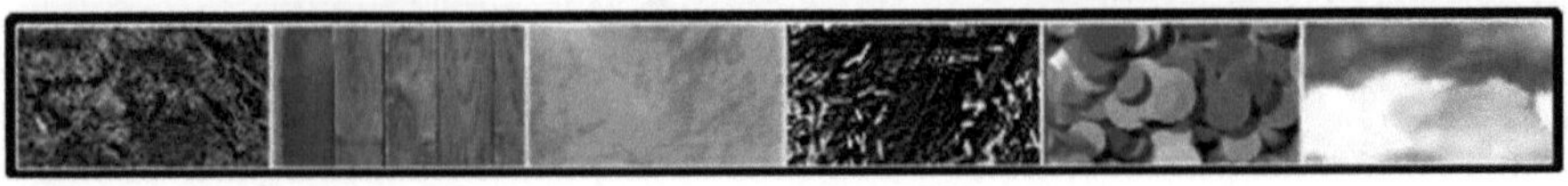

Fig. 5.13 Texture

5.1.3.1 Relationship to 3D Form

As previously discussed in the Sect. 5.1.2, the shape of a 3D object can partially be described by the shape of its silhouette. However, this can vary dramatically depending on how the object is perceived from various angles. Highly recognizable shapes are core components with 3D design. Often, just by silhouette alone, an object with a strong, distinctly recognizable shape can be recognized. In Gestalt psychology, this property is known as **Invariance**. When designing distinctive 3D objects whether they are commercial products or character designs, one step in checking that design is to simplify the 3D object down to a solo silhouette shape, by turning the object completely black and the background completely white. This is a recommended test analysis technique, especially with character designs. Most cartoon characters are easily identified by their silhouette.

3D designers have the potential to combine the principles of reification and invariance by modeling objects which look one way from one angle and/or have multistable negative spaces which appear within the object.

Shapes can be separated into rounded (curvilinear) or angular (rectilinear) forms. Rounded forms often represent gentle or kind emotions, whereas pointed angular forms can represent exiting or negative emotions. This connection between the forms extends beyond the visual to audio. Successful design often attempts the greatest engagement by trying to elicit multiple senses and emotions. There is some crossover and research between images and other senses, and this synesthesia (combining sensory perceptions) has been determined to be universal: "Speakers cross-linguistically associate non-words that have round vowels, such as/buba/, with round shapes, and non-words without round vowels, such as/kike/, with spiky shapes" [3] (Fig. 5.13).

5.1.4 Texture/Pattern

Textures describes the surface qualities of an object. This can cover large dichotomies of qualities: See Fig. 5.13 for some examples smooth/rough, wet/dry, shiny/dull, warm/cold, etc. All objects have some texture. In digital 3D design, these qualities are visually simulated. A "shader," which often models qualities of the real world, is given to the software which then interprets how the texture should be represented based on the angle of the camera to the source(s) of light in the scene.

In traditional design, the texture of an object can also be touched. This is not simply in sculpture, but the surface of a painting can also contain various textures. Vincent van Gogh was famous for building up paint into rough peaks on the surface of his canvas, a technique known as "impasto." Other techniques that draw attention to the texture of the object might be adding sand to paint or the use of various types of materials in the construction of a (collage) image.

Textures can be categorized in several ways:

5.1.4.1 Man-Made

These textures are often related in some way to the function of the object as well as the manipulation of both natural and purely constructed elements. Man-made textures demonstrate a connection to humans in one or more ways. Often the texture is connected to the construction process, for example, wood that has been lathed to form components of a chair. The construction process for lathing would aligns the grains of the wood to flow along the length of the chair which results in maximum strength of the constructed component. The grains of wood flow along the length of the chair leg (see Fig. 5.14). Man-made textures also reflect the purpose of an object, such as the reflective qualities of a street sign and the semiotic elements included on the surface.

5.1.4.2 Natural

This kind of texture tends to reflect the "organic" look and feel of nature. Specifically, this means they often reflect a greater connection and response to elements such as water, air, and sun.

Patterns:(see also Sect. 5.2.8)

They define an underlying structure ordering the texture. They exist in both natural and man-made textures.

5.1.4.3 3D Computer Graphics, Textures, and Connection to Mathematics

3D computer graphics are based on mathematics. Simulating 3D form and texture is a major area of research. The surface texture (especially shallow bumpiness) of an object within 3D graphics is often not generated by tessellating simulated form; it is often generated by a bump or normal map (see Fig. 5.15). This can be an efficient way to simulate visual complexity with increasing the number of polygons in a surface.

Our eyes are trained over time to translate and perceive a series of patterns into meaningful information. As a culture, we are now training our computers and

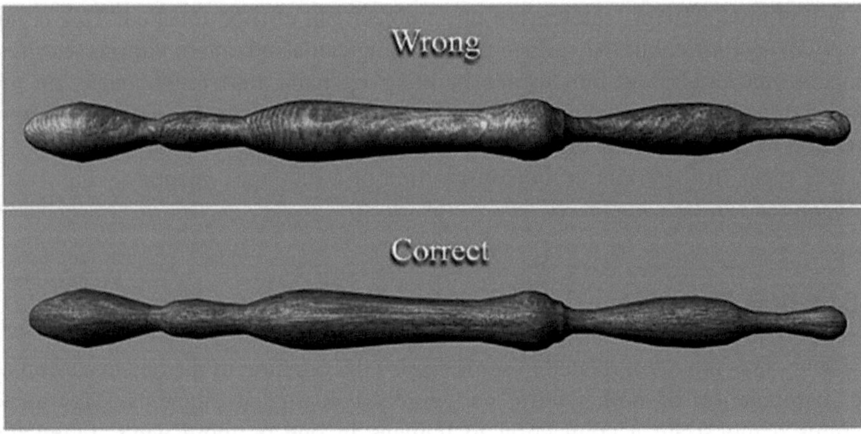

Fig. 5.14 Connecting design to reality

machines to also perceive those images and translate them into three-dimensional forms.

Numerous mathematicians have noticed the underlying mathematical nature of natural patterns [4]. This connection has been discussed by scientists for thousands of years (Cook). Many of the earliest recorded observations of math within natural patterns to spirals (see Figs. 5.16 and 5.17). Fibonacci, D'Arcy Thomas, and Alan Turing have described natural, visual phenomena which describe complex textures which all can be described using mathematics.

Many of the nature-inspired mathematical formulas have been applied to computer graphics texture creation. Some of the more common applications are "fractal generators" which use a repeating equation to generate an image. The Mandelbrot and Julia fractal equation sets are probably the most well known and common fractal types; however, there are many further fractals and combinations which have been identified.

Texturing 3D objects can be done with a mix of raster-based imagery. This is often photographic but sometimes created by the artist in an application like Photoshop. However, large photograph-based imagery used for textures can be inefficient and lead to unnecessary increased resource usage. Being able to mathematically generate natural textures has given artists increased flexibility to alter the textural pattern, increase the visual density, or recalculate for higher resolution. An example of this is with Perlin noise [5] (See Fig. 5.16).

Ken Perlin invented "Perlin noise" in 1982 and more recently simplex noise in 2001, which are algorithms for generating efficient math-based textures for efficient use in graphics applications. What is visualized is a simplified type of gradient noise, a complex, organic looking pattern, but at very low computer resource costs [1].

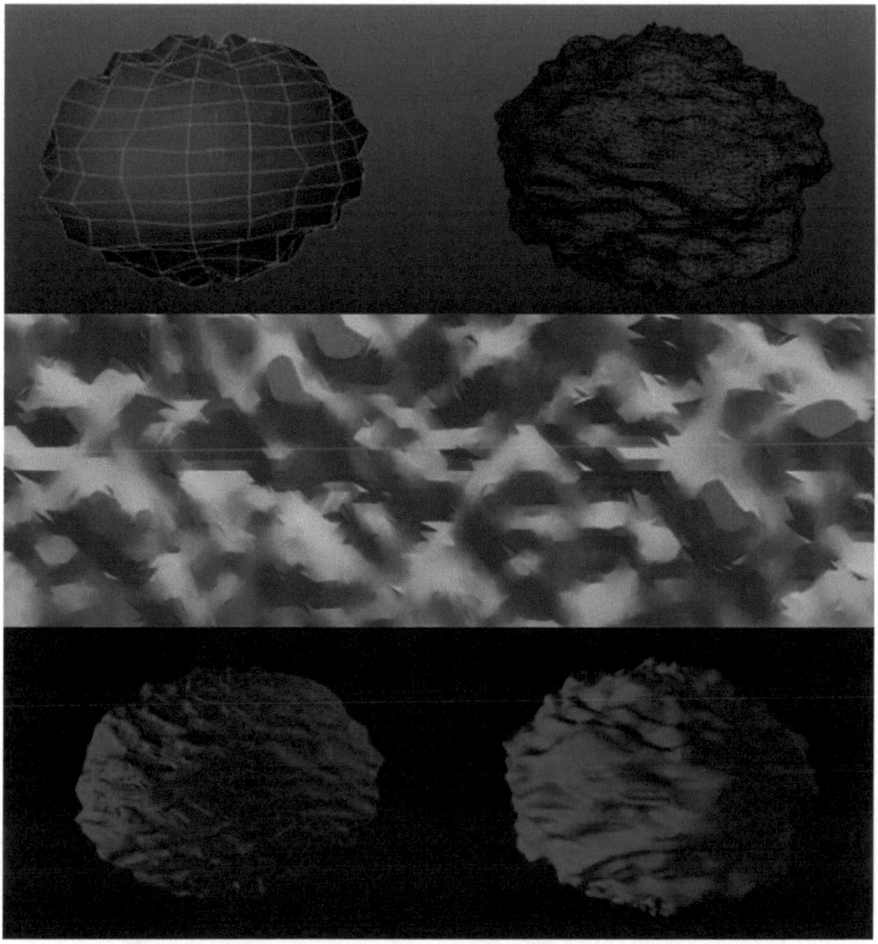

Fig. 5.15 400 faces versus 49,000 faces using a normal map

The connection of mathematics relates both to textures and in the form of objects. There are many programs that use algorithms to generate 3D objects. There are many natural objects which have tree-like branching structures which can be represented by fractals using computer-assisted algorithms for efficiency as an approach for generating geometry is rapidly being adopted. This rule-based approach to generating geometry rapidly combining with computer-vision- and photogrammetry-based approaches. While the current set of photogrammetry-based geometry creation and retopologizing tools and workflows being discussed in this book are human assisted, there is already work in using algorithms to further automate is foreseeable that many of these.

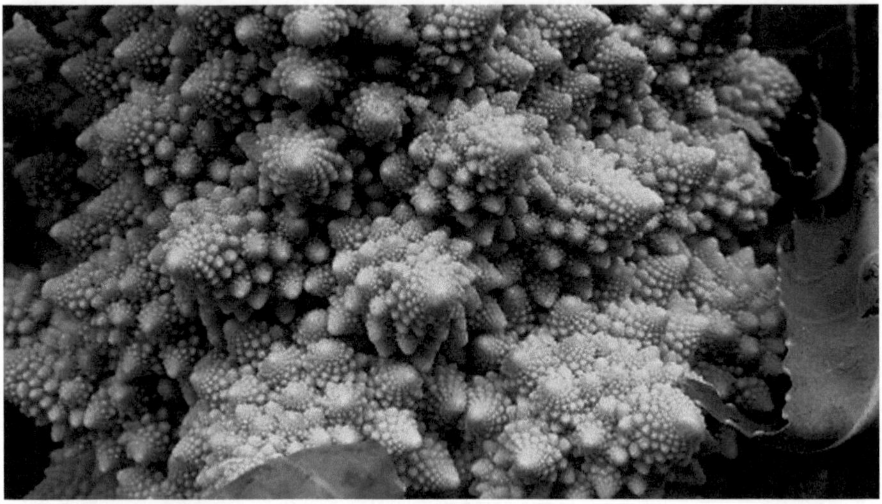

Fig. 5.16 *Romanesco Broccoli*

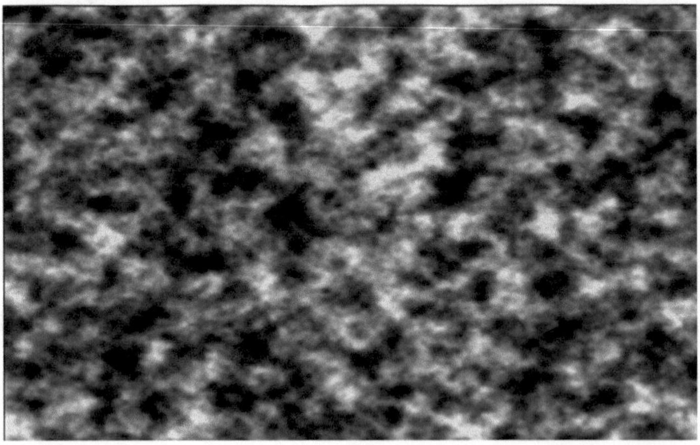

Fig. 5.17 Perlin noise

5.1.4.4 Texture Tools

There are an overwhelming number of tools for generating textures and applying them specifically for 3D computer graphics. Choosing the correct texturing application for a specific task is usually done by finding the tools and feature sets which most rapidly accomplish the task and within a budget. In addition to well-known programs such as Photoshop and internal 3D shading capabilities built into 3D programs such as Autodesk Maya (Hypershade), Mari (developed by The

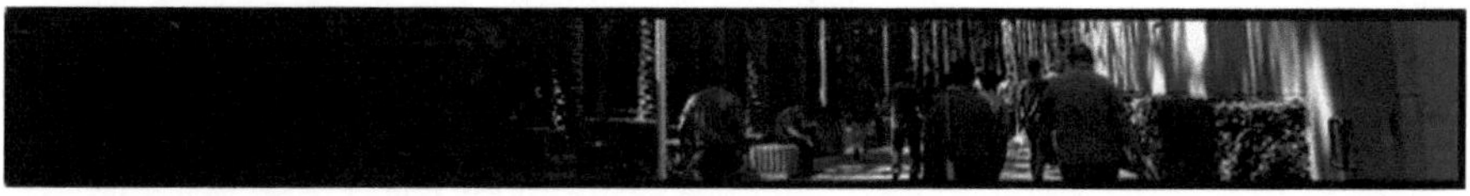

Fig. 5.18 Space

Foundry) and Substance Designer (developed by Algorithmic) are currently being heavily used by film and games industry, respectively (Fig. 5.18).

5.1.5 Space

Space separates visual elements within an image. However, how space is perceived within a design strongly influences overall feeling of created visuals. Regardless of 2D or 3D design, the use of space in an image is designed to engage the viewer. The way space is treated between 2D imageries relies on creating relationships between objects on a flat 2D image (X, Y coordinate system) or in simulating 3D imagery. Figure 5.18 shows an example of space represented using the context of humans and an environment.

How empty (or negative) space is designed also can be used to deliver additional information about an image. The use of space in an image can create a variety of emotional tones. For example, large gaps of space can create feelings of loneliness or very little space can create a feeling of confinement. Overlapping elements can help create visual engagement between objects, helping to make the viewer question the relationship between the overlapping objects. The way space is used can also help create rhythm or balance (see Sect. 5.2).

5.1.5.1 Relationship to 3D Form

The techniques of one-, two-, and three-point perspective used to generate within a 2D picture plane the perception of 3D space have been used for thousands of years. 3D computer graphics integrate true depth (Z) space within imagery. Whether done within 2D space or true 3D space "depth cues" can create or intensify feeling of depth within an image.

Depth Cues (see Fig. 5.19)

Color and shading: As objects recede, the color and details are reduced relative size: Objects "near" are perceived as being larger.

Overlap: Objects can be perceived as being in front of or behind based on overlap.

Linear perspective parallax: Lines that are parallel show convergence, if the imagery shows movement—objects in the foreground move more quickly than the background, known as perspective parallax.

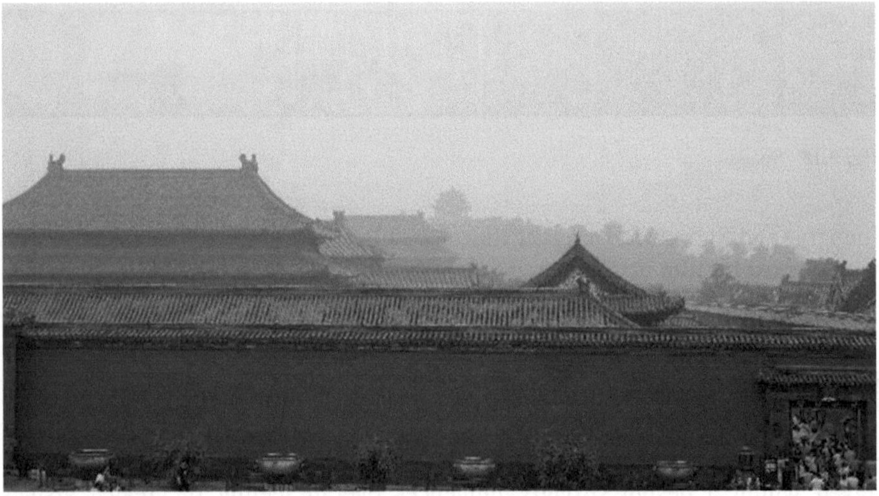

Fig. 5.19 Depth cues

Atmospheric perspective: As objects recede into the distance, they are reduced in visibility due to water vapor in the air.

3D computer graphics technologies generate images containing 3D spaces, often very similar to images created using moving or still-picture cameras. Later in this chapter, in the Sect. 5.2, we will discuss some of the techniques and conventions from the fields of photography and film cinematography that can be applied to 3D computer-generated imagery.

5.1.6 Form

This book specifically deals with the 3D form. Form as a design element is any object that can be measured with height and depth. A 3D object is understood by how it responds to various sources of light. For example, in Fig. 5.19, the sphere without light and dark would simply look like a (shape) circle. Additional qualities such as surface texture and bumpiness, specular highlights, reflectivity, refraction, and transparency further give visual clues about the form of an object.

There is significant overlap in the discussion of 2D and 3D forms. 3D objects can be constructed in a curvilinear, rectilinear, or a combination of the two. They can also have constructed or organic feeling to their form.

5.1.6.1 Types of Form

Forms can be categorized in multiple ways: how they fill physical space (**mass**) or how they surround empty space (**volume**). The mix of positive and negative spaces is a core component of defining the composition of 3D forms. They can also be categorized by 3D shape, for example organic, geometric, or mechanical forms.

5.1.6.2 Form Within Environment

Since 3D forms exist within 3D space, there are additional relationships that are created within the environments the forms are placed.

5.1.6.3 Form Tools

In this book, we will spend a great deal of time discussing how we achieve and develop form.

3D computer graphics comprehensive toolsets have the ability to achieve nearly any type of 3D forms. As discussed in Chapter 4, many of the tools allow for a variety of possibilities including a combination of hand-implemented sculptural, pushing rectilinear as well as procedurally generated naturalistic sculpting as well. As discussed in Chap. 2, the nature of technology often causes significant convergence among fields. This is happening within the fields of design field as well. When the field of 3D graphics was new simply knowing basic 3D modeling was often enough to obtain gainful employment. Growth in the number of 3D modeling specialists in industry combined with photogrammetry and other accelerated workflows have reduced demand for 3D modeling jobs. In addition, as discussed in Chap. 2, the changes in accelerating technology become disruptive.
As 3D asset creation becomes more automated, additional toolsets in design and cinematography will become ways to differentiate 3D artists from technicians.

5.2 Principles of Design

Principles of design are used to provide organization and structure to the elements of design. Within the design field, there is no exact list of principles of design. However, there is a fairly consistent core group. Due to the nature of this book as a discussion of a convergence between multiple technologies and fields of design, we will also address some of the design components from the fields of photography and moving picture cinematography as well as the following five principles of design:

unity,
emphasis,
balance,
repetition, and
dynamics.

Within this list, we will also cover several subcategories. Additionally, while in other design books, principles of design often cover both 2D design and maybe 3D design [6]. For the purpose of this book, we are going to specifically focus on the application of these principles with the understanding that the final image is one that is created using 3D forms.

Designing 3D computer models, from computer-vision-accelerated stereophotogrammetry as discussed in previous chapters, uses knowledge from multiple fields. The convergent fields and technologies we will discuss in the next few pages are photography, film, and animation and Gestalt psychology.

5.2.1 Camera: Lens, Film, and Exposure

5.2.1.1 Lens

The medium of image creation greatly affects what is perceived in the final design. Whether a designer modifies perspective artistically or a cinematographer uses various lenses, how depth is perceived greatly affects the final composition. Two images taken from the same location with different lenses generate significantly different visual output.

A long lens (telephoto, 70–300 m+) will significantly compress the perception of depth within an image. In addition to this quality, the length of the lens also affects the plane of focus, which is the distance from the lens where objects are in focus.

A short lens (wide angle <21 mm) will deliver a wider angle of view. Wide-angle lenses accentuate the feeling of depth by warping an image. All types of lenses can be simulated in order to achieve appropriate "feeling" for the image within 3D digital graphics.

Depth of field specifies what objects are in focus and out of focus. A shallow focus can help generate images and designs which guide the viewer to a desired object. With real-world optics, it is sometimes difficult to generate specific sharpness of a subject with exact blur for the out-of-focus areas. Within 3D computer graphics, users can generate an additional Z (or depth) channel within many file types allowing the creator to specify depth of field as part of the post-production process. In traditional film, this was impossible to accomplish. The Lytro camera, using light field technology, is now able to capture light and depth in a way that allows post-process planes of focus. In addition to the Lytro camera, within the photogrammetry process, the camera, lens, and focal distance information are becoming standard components of metadata within calculations used to generate photogrammetry output. As additional cameras adopt depth-sensing technologies (such as the PrimeSense depth sensor on the Kinect), this will give further accuracy and interactivity to combining image captures in the real word as both photographs and 3D objects.

5.2.1.2 Film and Exposure

While photochemical film is largely a technology of the past, there are significant parallels to the visual output generated by 3D-rendering engines to film stocks of the past. One characteristic that parallels film type is the renderer. Each 3D computer graphics program's render has default "looks" that are generated. These default looks can be modified. Factors similar to the speed of film (how fast it captured light) dictated how quick the shutter needed to open and close as well as the width of the aperture of the camera in order to expose the film. The resulting image would be further affected by how much grain was visible within the image as well as motion blur. Within 3D graphics, motion blur and exposure are highly controllable elements that can be specified within the software or also done as part of the post-production process.

5.2.2 Cinematography

While many of the principles of design that will be discussed below deal with elements of **composition**, in cinematography, composition refers to what is captured by the camera in frame of the image. The **mise-en-scène** (from the French) "on the stage," encompasses everything from actors, props, lighting, and even the type of lens used on camera. While there is significant intentionality to the cinematography, however, there is potentially greater organic latitude due to the moving and potentially unpredictable nature of the elements within the frame. In 3D computer graphics, almost all elements are currently planned. Because of this, we will mostly focus on the still image within cinematography. We will separate framing into the categories dynamic and static.

5.2.2.1 Dynamic Composition

In our previous discussion on the elements of design, the use of line greatly affects the emotional impact of an image. When an image has strong diagonal lines, the diagonals in an image create a feeling of excitement. Lines also guide the eye around the image. The composition of an image with multiple triangular shapes helps create visual tension or excitement, whereas an image with multiple curved shapes can have a gentle or sensual.

How the objects in the image sit within the frame of the image can create dynamic or static composition. We will discuss this further in the sections below on (2) 5.2.6 and (8) 5.2.5.

5.2.2.2 Static Composition

Strong horizontal or vertical lines are often considered static, calm, or soothing. When the objects in the scene are also centered in the frame, this also creates a static composition. The vertical or horizontal lines can be generated by man-made or natural elements.

5.2.3 Unity

It is often the first and/or final consideration of any design. To achieve unity, a variety of elements need to work together to create a cohesive image. This design principle is connected to and supported by all of the other elements and principles of design. It is also conceptually close to the Gestalt principles of design.

This concept deals with connecting all of the elements in the image into a cohesive whole in a way that all should work together to support and amplify the core idea or emotion being expressed.

An example of this would be an image that has a relaxed or gentle subject that could be further unified by additional round shapes, smooth textures, and reduced total contrast. Reversing this concept for an image that has a dynamic, exciting feeling could be further unified by adding additional diagonals, triangular shapes, rough textures, and high tonal contrast. Applying consistent post-process “on top” an image can help to unify it: grain, color, blur, and other post-processing on top of the image.

Additional ways to increase unity in images as well as image sequences would be to stage the image in a way that would visually link the imagery in a way that would guide the eye around the image.

Unity is also highly related to design philosophy, physiology, and psychology of Gestalt. The word Gestalt means that the organized whole is perceived in a way that is greater than the sum of its parts. The human brain naturally looks for ways to unify imagery [7]. Much of the Gestalt philosophy deals with how the brain groups and categorizes visual imagery. Additional Gestalt principles can be used to help unify elements if an image needs additional help. These are continuity, alignment and containment, closure, proximity, and similarity.

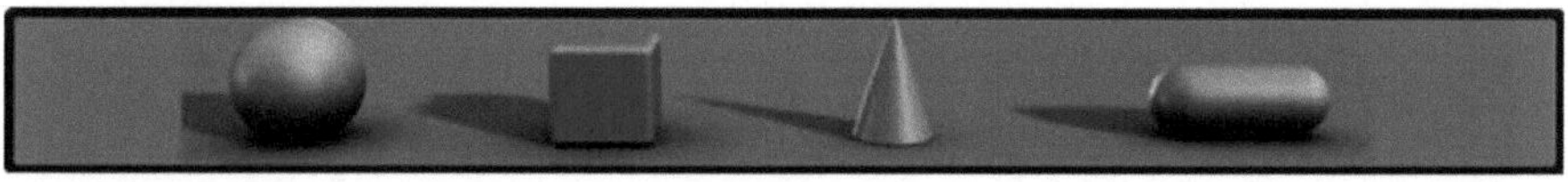

Fig. 5.20 Forms

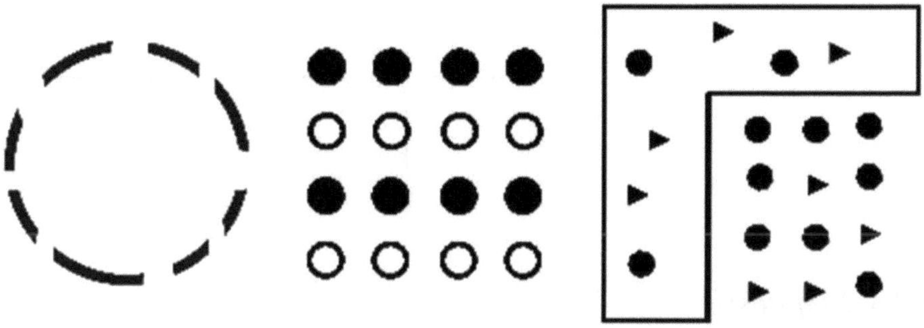

Fig. 5.21 Gestalt grouping principles

5.2.4 Gestalt Grouping

A major focus of Gestalt psychology is based on the implication that the mind translates external stimuli holistically rather than the sum of their parts.

Continuity: Objects that overlap in 3D space are often connected by the brain.

Alignment and Containment: It is also known as uniform connectedness [8]. Objects that are arranged within 3D space in ways that put them on similar planes of height make them. Surrounding the objects within the rectangular form as shown in the left Fig. 5.20 helps to group them, even though they are not even the same shape.

Closure: The brain has a natural tendency to connect (see also proximity) or close a space—do you see a circle in Fig. 5.20 left.

Proximity: Objects that are near each other are connected by the brain into groups (Fig. 5.20 center).

Similarity: Components within an assortment of objects are cognitively grouped if they are similar to each other. Making exact repetitions of 3D objects is very simple in 3D computer graphics, making similarity and **repetition** (see also a greater description of this design principle) to be strong unifying components (Fig. 5.21 right) [9].

Fig. 5.22 Rule of thirds

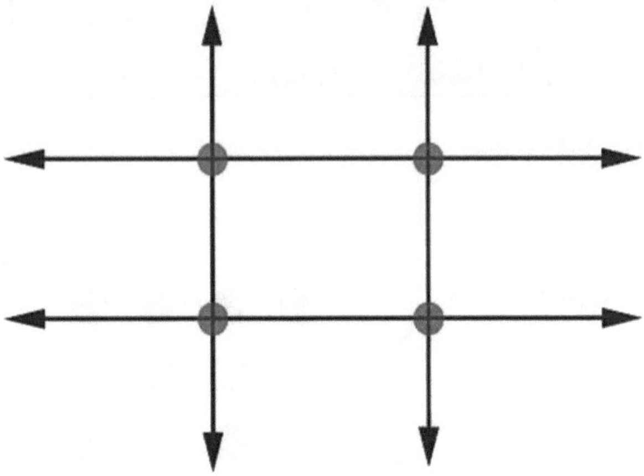

5.2.5 *Emphasis*

It refers to the point of focus or interruption. It indicates the locations in an image, which most strongly attracts the viewer's attention. There are usually a primary and sometimes secondary areas of emphasis. Emphasis can be achieved in many ways. One is to create outstanding elements by contrasting form, luminance, or color, to create a break in rhythm or dynamics, or dominance within the visual hierarchy (example, increased scale within a set of multiple forms).

5.2.5.1 Placement

The rule of thirds divides an image into thirds on the horizontal and vertical axes (Fig. 5.21). Placement at one of the corners of intersection is one area where there is emphasis. In addition, throwing the image of center, when used appropriately, can create additional dynamic imagery with emphasis placed on areas of the image, which further strengthens the composition (Fig. 5.22).

Focal point—what is the center of design. What is visually dominating?

In addition to how form is placed and fills the visual field, there are many additional ways to add differentiation.

5.2.5.2 Size and Proportion

Proportion deals with the relationship between objects. The relative size of objects within a composition helps to inform size and spatial relationships between objects. How their relative size and scale is imaged in the context of the other objects in the scene works to develop the visual hierarchy. Other depth cues may be used (see Sect. 5.1), but the standard cues for size within an image are the ones best known to us; humans. 3D visual designs can proportionally show the human form as huge or small against the background, or even other human forms. This effectively generates emphasis on either the human or the environment, as well as informing the viewer as to the size of the environment.

5.2.5.3 Contrast

The brain is specifically designed to notice differences. Using various methods to achieve visual or conceptual contrast helps to engage the viewer. Creating contrast within an image can be done in many ways. Each of the elements of design (color, line, shape, value, texture, and form) can be used in ways that create contrast.

Using opposing or juxtaposing elements within the design can create contrast. This can be specifically done with in one or several elements of design such as:

Color: Use complimentary colors (colors that are opposite from each other on the color wheel)

Tone: Perhaps the most obvious is creating areas of strong brightness and darkness.

Shapeand Line: contrast curvilinear and rectilinear forms, or smooth and jagged lines.

The main contrast in an image should usually be located around the center of interest. Having too much contrast (variety) in all areas can destroy unity as well as create a feeling of chaos and confusion (unless this is the goal of the image).

5.2.6 Balance: Symmetry and Asymmetry

"Balance is a state of equalized tension and equilibrium, which may not always be calm" (White). As a design principle, balance is a visual concept which stems from the physical world, where an object must reconcile between opposing physical forces such as gravity, slope of the surface it sits on, shape of the object, and that object's center of gravity in order to achieve balance or tip over.

For a digital image of a 3D object, balance within that object can be achieved with symmetry, with asymmetry, or with radial balance. How the 3D object is realized as an image and achieves balance will also depend on how it sits within the context of the image frame as well how the object is lit.

5.2.6.1 Symmetry

Similar to the same principle in physics, imagery can be defined in ways where an image's visual weight can be imagined. Symmetrical organization is created through the alignment of graphics along an axis. The image plane can be imagined as a scale, and the focal point of the image is the fulcrum (Fig. 5.23). Large

Fig. 5.23 Symmetry

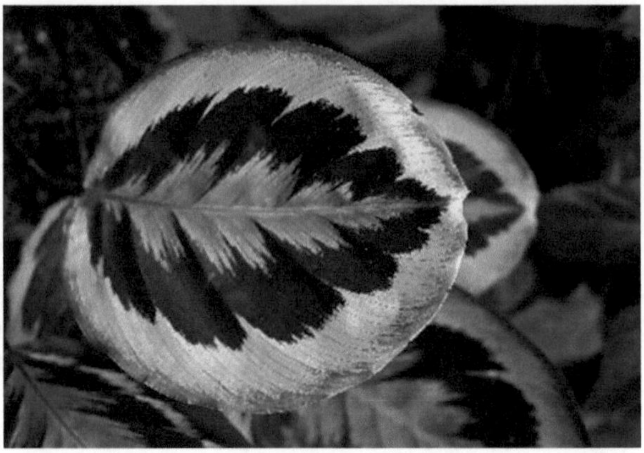

Fig. 5.24 Asymmetry balance

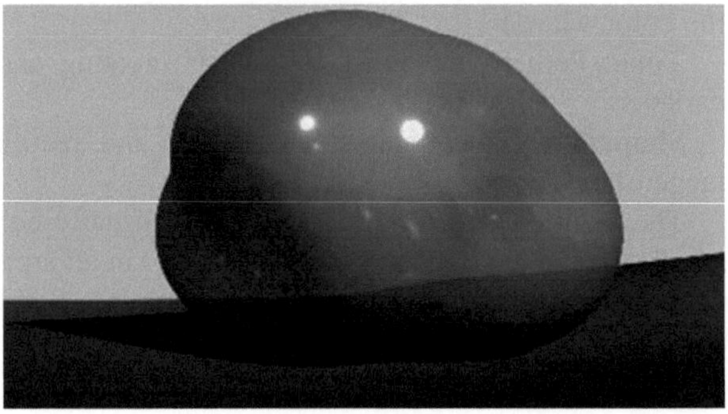

symmetrical shapes when placed in the center of an image are considered in balance. The weight of each shape within an image can also vary, depending on how it is visualized. Light colors, transparency, and little shadows can reduce the weight of an object in an image (Fig. 5.24).

5.2.6.2 Asymmetry

Also known as **Informal Balance**, **asymmetry** is conceptually more complex and deals with creating balance within a composition by distributing visual weight among elements around a fulcrum point. This can be done by placing imagery of various visual "mass" on different sides of the fulcrum point. Figure 5.24 is asymmetric, but because its center of mass is distributed, the object is centered on the frame and appears to be resting on a surface and this creates a feeling of balance within the image.

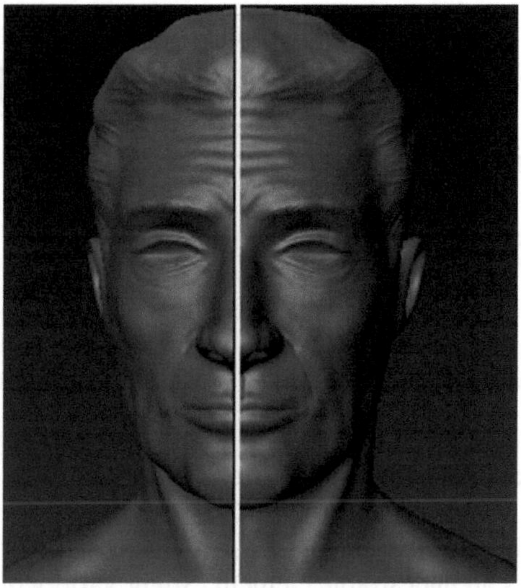

Fig. 5.25 Facial symmetry

5.2.6.3 Symmetry Tools

There are many tools available for aligning objects both within the context of the image plane and within the context of other objects. Many modeling and sculpting workflows now include the ability to create objects symmetrically, this can greatly accelerate initial designs. This can be done in one of the two ways. One way is to start by building half of the object and creating a mirror image duplicate of itself. The second way is that many programs now often include a symmetry button for creating objects. As an example (Fig. 5.25), the face was modeled with "symmetry on" every detail sculpted on one side translated to the other. In the real world, faces are rarely 100 % symmetrical; however, it is a factor that connects to repetition as well as aesthetic beauty [10]. Working using the second approach works better because of the ability to visually check the final result.

5.2.7 Repetition

Repetition with variation is often used within design. Repetition can help unify an object, by creating a series of repeating forms. How those forms are laid out connects repetition to the principle of rhythm (see Discussion in Sect. 5.2.8). However, without any variation, the repetition can become boring. One reason for boredom is that the brain can too easily and immediately perceive the image in a single glance. However, when variety is added within repetition, it creates a stronger visual image which will draw the eye in and engage the brain in a way that will cause further consideration.

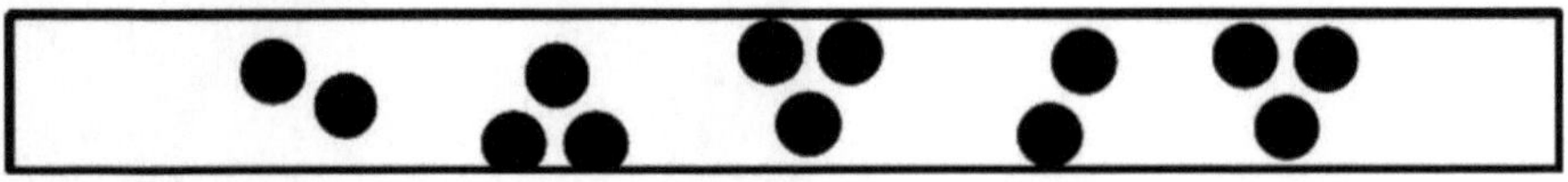

Fig. 5.26 Proximity

5.2.7.1 Repetition Tools

In most programs, there is the ability to instantly create an array of objects. When generating the array, the user can specify how many objects are created, the distance between objects as well as additional rotation (etc.).

5.2.7.2 Variety Within Repetition

For creating 3D objects with naturalistic or organic feeling, exact duplication often is an undesirable element. Slight variety of rotation, scale, or color can help throw off the eye from noticing the repetition.

5.2.8 Proximity

This principle can be applied in 2D and 3D space designs as well temporal formats, such as film and motion graphics. How objects are placed next to each other in space and time affects our perception of them as a connected unit. When several objects are placed close to each other as shown in Fig. 5.26 above, the brain easily groups them into 5 units. This principle can be connected in a temporal sense to [11], which takes the images projected next to each other within a time proximity (between cuts in a film) and creates an association between them. The context and order that the images are perceived can have a great impact on the emotional impact (Isenhour).

Visual hierarchy is the order in which humans perceive an image. This order is created by several factors. Objects with high visual contrast to their surroundings are perceived first. Visual contrast can be achieved by structuring the elements of design in ways that make objects stand out. Some examples of objects that are designed to stand out are warning signs and tennis balls. The bright fluorescent colors catch the eye. Opposite to this would be a soldier wearing camouflage, which has the purpose of blending and hiding the person into the background.

5.2.8.1 Proximity Tools

Spot Healing Brush: This Photoshop tool analyzes surrounding pixels to check for similarity within surrounding colors, with the new "content-aware" technology, and it looks for "proximity of pattern" in the surrounding area and will attempt to create additional patterns.

5.2.8.2 Hierarchy Tools

Using a combination of tools related to the elements of design can help describe hierarchy:

Scale: The larger the object, the more dominant it can become within the visual hierarchy.

Contrast: Making the object stand out by contrasting it to what is surrounding it.

Position: Where the object is represented in the frame also can affect hierarchy. The top of the frame increases importance.

5.2.9 Dynamics

This principle of design is very strongly connected to both audio and movement. The feeling an image gives is usually related to the use of patterns and line.

5.2.9.1 Direction and Movement (Similar to Principle of Staging)

This can be accomplished by the placement of the 3D object so that the profile creates lines that are diagonal to the frame of the image. Lines can also be a part of the texture that covers the 3D object creating additional or contrary feelings of direction or movement within an object.

5.2.9.2 Rhythm

Visual rhythm can be created in several ways. Individual lines can create a linear rhythm. How lines are placed on a page affects the pace of how they are perceived by the human eye. A strong use of repetition with any of the elements of design can create a feeling of rhythm within an image. An example of a very simple rhythm would be a repetition of a singular theme, and using alternation of a differentiated sequence of repeating motifs can generate a more textured and complex rhythm.

5.2.9.3 Arrhythm

Elements presented arrhythmically (without rhythm) can give contrast to areas with visual rhythm. The word specifically means "against rhythm." Arrhythmic elements can be used effectively in designs to create emotional feelings which connect to disharmony.

5.2.9.4 Random

Arrhythm taken to an extreme becomes random. Depending on the size of the random elements, this can be effective as a naturalistic visual element.

References

1. Albers, J.: Interaction of color: Josef Albers. Yale University Press, New Haven. Also Interaction of Color: 50th Anniversary (app) (2006)
2. Motter, B.C., Belky, E.J.: The guidance of eye movements during active visual search. Vision. Res. **38**(12), 1805–1815 (1998)
3. D'Onofrio, A.: Refining bouba-kiki: Phonetic detail and object dimensionality in sound-shape correspondences. J. Acoust. Soc. Am. **132**(3), 1968 (2012)
4. Cook, T.A: The Curves of Life: Being an Account of Spiral Formations and Their Applications to Growth in Nature, to Science and to Art with Special Reference to the Ms. of Leonardo Da Vinci; with 11 Plates. Courier Dover Publications, Mineola (1979)
5. Perlin, K.: Improving noise. ACM Trans. Graph. (TOG) **21**(3), 681–682 (2002)
6. Stewart, M.: Launching the Imagination: A Comprehensive Guide to Three-Dimensional Design. McGraw-Hill, New York (2006)
7. Palmer, S., Rock, I.: Rethinking perceptual organization: The role of uniform connectedness. Psychon. Bull. Rev. **1**(1), 29–55 (1994)
8. Lidwell, W., Holden, K., Butler, J.: Universal Principles of Design: 125 Ways to Enhance Usability, Influence Perception, Increase Appeal, Make Better Design Decisions, and Tech Through Design [25 Additional Design Principles]. Rockport publishers (2010)
9. White, A.W.: The elements of graphic design. Skyhorse Publishing Inc, New York (2011)
10. Thornhill, R., Gangestad, S.W.: Human facial beauty. Hum. Nat. **4**(3), 237–269 (1993)
11. Isenhour, J.P.: The effects of context and order in film editing. Education Tech. Research Dev. **23**(1), 69–80 (1975)

Chapter 6
Conclusions and Looking Forward

Abstract This book concludes by looking at additional market leaders and how they are further integrating technology for workflows as well as adapting and converging.

Keywords Conclusions

As we have shown in this book, the development of 3D computer modeling workflows has undergone substantial changes in its short history. At any point in this development cycle, a case can be made that a "paradigm shift" has taken place—the transition from NURBS to polygon modeling, for example, or the ability to represent fine detail with normal maps.

The integration of photogrammetry, high-polygon-density interactive modeling, and retopology represents a sweeping change that some might say will completely overturn traditional approaches. They have certainly represented a major change in approach for many large studios. As the tools for this workflow have proliferated and become more accessible to more people at a lower price point, the process has become the norm for an increasing number of smaller studios, independent artists, and education institutions.

As this change in tools happens (as discussed in the chapter on the nature of technology), adoption speed of new tools is also accelerating. As these new technologies and tools become part of the status quo, we anticipate greater fidelity within our 3D models and almost automatic retopologization within the next few years to be ubiquitous.

Looking to Entertainment VFX market leader Weta Digital, over the last two years, Weta upgraded the photogrammetry pipeline and added high-fidelity LIDAR (laser scanning). For the new Hobbit films, they record "much more… total data than ever before… for example, Avatar was approximately one petabyte, now it is over six times that for the Hobbit." They have reached the point where they are "scanning every set that they are shooting on" using a Leica ScanStation C10, and they can scan 50 million points per scan in 30 min, and about 10× scans per set, plus high dynamic rage photographs.

S. Foster and D. Halbstein, *Integrating 3D Modeling, Photogrammetry and Design*, SpringerBriefs in Computer Science, DOI: 10.1007/978-1-4471-6329-9_6,

As the technology for handling this quantity of data reaches consumer price points, this opens up further design possibilities. Higher-fidelity visual 3D scans from real life; using design principles to sculpt and retopologize output in a multitude of forms will become standard and semiautomated workflows.

With rapid optimization and automation of 3D scanning and modeling, we anticipate further convergence among technical and design fields. For example, combining scanning with motion capture, augmented reality, and visual effects [1], 3D models will be combined with future generations of tools for designers to create with.

Reference

1. Seymour, M.: The hobbit: Weta returns to middle-earth, FXGuide http://www.fxguide.com/featured/the-hobbit-weta/(December 2012)

Chapter 7
Resources and Application

Abstract Resources chapter—links for photogrammetry and design.

Keywords Resources · Photogrammetry · 3D scanning · Color

7.1 PhotogrammetryResources

123D App Autodesk
http://www.123dapp.com/
Arc3D
http://www.arc3d.be/
Aero Scan
http://areo.co.nz/
Hypr3D
http://www.hypr3d.com/
My3D Scanner
http://www.my3dscanner.com/
Microsoft Kinect SDK
http://www.microsoft.com/en-us/kinectforwindows/
Microsoft Photosynth
http://photosynth.net/
To download the Photosynth application:
http://cdn1.ps1.photosynth.net/installer/2013-02-27/PhotosynthInstall.exe
Insight 3D
http://insight3d.sourceforge.net/
International Society for Photogrammetryand Remote Sensing
http://www.isprs.org/
OSM Bundler
https://code.google.com/p/osm-bundler/

S. Foster and D. Halbstein, *Integrating 3D Modeling, Photogrammetry and Design*,
SpringerBriefs in Computer Science, DOI: 10.1007/978-1-4471-6329-9_7,

PhotoModeler Software
http://www.photomodeler.com/index.html
ReconstructMe—Microsoft Kinect for 3D digitizing of objects
http://reconstructme.net/
VisualSFM: University of Seattle
http://homes.cs.washington.edu/~ccwu/vsfm/
CMP SfM Web Service (Europe)
http://ptak.felk.cvut.cz/sfmservice/
Structure from Motion Toolkit
http://www.visual-experiments.com/demos/sfmtoolkit/
Visual Size
http://visualsize.com/
Good Forums
Photosynth Rivals
http://pgrammetry.com/forum/
Other Application:
PhotoCity (game)
http://photocitygame.com/
3D Scanningtechnologies
http://www.applications3d.com/icam.html?gclid=CNO4neLpl7YCFalxOgodpE0Anw
CyArk: Digital preservation of world heritage sites
http://archive.cyark.org/

7.2 Retopologizing Resources

Autodesk Maya (2014)
http://www.autodesk.com/products/autodesk-maya/overview
Advanced Modeling Workflow: Retopo tools: Quad-draw, shrink wrap
http://www.youtube.com/watch?v=DgjioOigaQA
Autodesk Mudbox
http://www.autodesk.com/products/mudbox/overview
3D Coat
http://3d-coat.com/
Topogun
www.topogun.com
Pixologic: ZBrush
http://pixologic.com/zbrush/features/QRemesher-retopology/

7.3 Color Resources

kuler.adobe.com

Albers, Joseph Interaction of Color (iPad app and book)

ColrD—Free Color Chrome App

References

1. Richards, J.A.: Remote sensing digital image analysis: An introduction (09/13/2012)(ISBN:3-642-30061-8, 978-3-642-30061-5)
2. http://www.youtube.com/watch?v=Cd6hnSJ-cQYYouTube: Photogrammetry history p2
3. Dyer: Mitch Xbox's Kinect one is legitimately awesome May 2013http://www.ign.com/articles/2013/05/22/xbox-ones-kinect-is-legitimately-awesome
4. Reeves, B-GameInformer: Infamous second son: How sucker punch photocopies pedestrians for infamous second son May 2013https://www.gameinformer.com/games/infamous_second_son/b/playstation4/archive/2013/05/24https://www.gameinformer.com/how-sucker-punch-photocopied-actors-for-infamous-second-son.aspx
5. Seymore, M.: The hobbit Weta returns to middle-earth, FXGuidehttp://www.fxguide.com/featured/the-hobbit-weta/
6. Zbrush Webinar–ZRemesher™ http://www.youtube.com/watch?v=lPCDOzep8NQ

Index

S. Foster and D. Halbstein, *Integrating 3D Modeling, Photogrammetry and Design*,
SpringerBriefs in Computer Science, DOI: 10.1007/978-1-4471-6329-9,

MIX
Papier aus verantwortungsvollen Quellen
Paper from responsible sources
FSC® C105338

If you have any concerns about our products,
you can contact us on
ProductSafety@springernature.com

In case Publisher is established outside the EU,
the EU authorized representative is:
Springer Nature Customer Service Center GmbH
Europaplatz 3, 69115 Heidelberg, Germany

Printed by Libri Plureos GmbH
in Hamburg, Germany